T0186012

GENDERED EXPERIENCES OF GENOCIDE

Voices in Development Management

Series Editor:
Margaret Grieco
Napier University, Scotland

The Voices in Development Management series provides a forum in which grass roots organisations and development practitioners can voice their views and present their perspectives along with the conventional development experts. Many of the volumes in the series will contain explicit debates between various voices in development and permit the suite of neglected development issues such as gender and transport or the microcredit needs of low income communities to receive appropriate public and professional attention.

Also in the series

Participatory Development in Kenya
Josephine Syokau Mwanzia and Robert Craig Strathdee
ISBN 978 0 7546 7877 9

Dot Com Mantra
Social Computing in the Central Himalayas
Payal Arora
ISBN 978 1 4094 0107 0

The Dominance of Management
A Participatory Critique
Leonard Holmes
ISBN 978 0 7546 1184 4

Losing Paradise
The Water Crisis in the Mediterranean
Edited by Gail Holst-Warhaft and Tammo Steenhuis
ISBN 978 0 7546 7573 0

Tourism, Development and Terrorism in Bali
Michael Hitchcock and I Nyoman Darma Putra
ISBN 978 0 7546 4866 6

Gendered Experiences of Genocide
Anfal Survivors in Kurdistan-Iraq

CHOMAN HARDI
St Antony's College, Oxford, UK

Routledge
Taylor & Francis Group

LONDON AND NEW YORK

First published 2011 by Ashgate Publishing

2 Park Square, Milton Park, Abingdon, Oxon OX14 4RN
711 Third Avenue, New York, NY 10017, USA

Routledge is an imprint of the Taylor & Francis Group, an informa business

First issued in paperback 2016

British Library Cataloguing in Publication Data
Gendered experiences of genocide : Anfal survivors in
   Kurdistan-Iraq. -- (Voices in development management)
   1. Anfal Campaign, Iraq, 1986-1989--Women. 2. Genocide--
   Iraq--Kurdistan--Sociological aspects. 3. Women,
   Kurdish--Crimes against--Iraq--Kurdistan. 4. Women,
   Kurdish--Iraq--Kurdistan--Social conditions.
   I. Series
   956.7'20441-dc22

Library of Congress Cataloging-in-Publication Data
Hardi, Choman.
   Gendered experiences of genocide : Anfal survivors in Kurdistan-Iraq / by Choman Hardi.
      p. cm. -- (Voices in development management)
   Includes bibliographical references and index.
   ISBN 978-0-7546-7715-4 (hardback)
   1. Women, Kurdish--Social conditions. 2. Genocide--Kurdistan--History. 3. Anfal
Campaign, Iraq, 1986-1989. 4. Iran-Iraq War, 1980-1988--Atrocities--Iraq. I. Title.

   HQ1726.5.H367 2010
   956.7'20441082--dc22

                                                                    2009051192

ISBN 978-0-7546-7715-4 (hbk)
ISBN 978-1-138-26029-0 (pbk)

Contents

Preface

Anfal started when I was in my early teens. I was living in the city of Suleimanya with my parents. My three sisters and three brothers had already left home. My eldest brother and sister lived in the West. My other brothers were peshmarga (freedom fighters) in the mountains. My second sister was married and lived in Tikreet (Saddam Hussein's birth place) and my third sister was finishing her medical degree in Mosul University. From February 1988, there were whispers about massive Iraqi strikes on the Kurdish countryside. There were talks of widespread gassing, bombing, military siege, demolition of villages, looting, and the capture and mass disappearance of civilian populations.

One night my father's cousin came to tell us that some of the villagers had been brought to the Emergency Forces building in Suleimanya. The place acted as one of the numerous temporary holding centres during Anfal. Hundreds of civilians could be seen beyond the fences of the building. They looked pitiful, with torn clothes that were drenched in mud. He told us that people from the surrounding neighbourhoods went to 'throw food at the prisoners.' I carried this image in my head for years; people trapped in animal cages desperate for food and water. My father predicted that they will all be killed but I refused to believe him. For most of my life I resisted my father's pessimism about the world. It was years before I could accept that people are capable of such systematic cruelty.

In April, as news of the devastation continued, a woman came to tell us that she was harbouring my brother, Asos. My brothers were peshmarga in the Qaradagh region which became prey to the second Anfal offensive. In the chaos of a swift defeat they were separated from each other and Asos had sneaked back into town with some villagers. He had thought it best not to come home. He knocked on the door of an old friend from university who lived in a small town outside the city. This family gave refuge to my brother until the curfew, and house by house search of Suleimanya, a few weeks later. After this he managed to get forged papers and go to Tikreet. It was thought that this would be the safest place for him to hide. After a few weeks we heard from my brother Rebin who, after being mildly injured in the gas attack on Shanakhse, had made it to Iran. Soon Asos joined him there.

In August 1988 when Iraq signed the peace treaty with Iran, more than two weeks before Anfal came to an end, my father decided to leave Iraq and join my brothers in Iran. We travelled by mule, crossing through the mountainous 'prohibited region', guided by smugglers. We passed through dozens of deserted villages. It is the eerie silence of the abandoned houses and farms that I remember the most.

While living in Iran I heard many stories about Anfal and, at the age of 14, I started taking notes about the campaign. I asked our friends and relatives, who had been in the midst of the operation, to write about their experiences in a notebook I dedicated for this purpose. This book was kept by a man, himself a survivor, and was never returned.

No one knew what exactly had happened to the captured civilians. Many people hoped that the villagers were either imprisoned or relocated to the Arab south which was the fashion in the sixties and seventies (Van Bruinessen 1988). Finally, during the short-lived 1991 popular uprising, the security and intelligence offices were raided in Iraqi Kurdistan and tons of documents were captured which revealed the truth about Anfal. The documents, alongside witness testimonies and mass grave sites, made it clear that the disappeared civilians had been executed in 1988.

In 1993, I arrived in England and launched into learning English and catching up with my education (I had missed two years of schooling due to the post-first Gulf War chaos). I now realise that for a number of years I subconsciously blocked out many traumatic issues concerning my home country, in order to manage the tasks required for my adaptation in the new country. I continued my education and read philosophy and psychology at Oxford University. I went on to do an MA in philosophy at University of London in 1999. In 2001 I was fortunate to secure a scholarship to do a PhD on the mental health of Kurdish women refugees at University of Kent in Canterbury.

It was when I was doing my MA that my interest in Anfal was sparked once again. I gradually lost interest in philosophy and the abstract questions it dealt with. I became more interested in why violence happens in certain communities and what happens in the aftermath. While living in a peaceful democracy, I became acutely aware of the brutalisation of my community, the increased violence against women, and the social inequalities that were widespread. I wanted to know more about how communities can recover from violence and what steps can be taken towards the attainment of justice and equality.

Later, when I was interpreting for refugees and asylum seekers I met large numbers of young Iraqi Kurds who were illiterate or semi literate. This was strange considering how the majority of people in the 1970s and 1980s were graduates. At times, I even wondered if some people were claiming to be illiterate because they were wrongly advised by agents and smugglers that this would be beneficial. Soon I realised that some of these were the Anfal surviving children who were born in the Kurdish villages before Anfal forever changed their lives.

I watched numerous documentaries about Anfal when the Kurdish satellite channels were launched in the late 1990s. These documentaries mainly consisted of interviews with survivors. They were like a window through which we heard the stories of a community brought to its knees by terror, death, torture and mass disappearance. As I listened to the men and women, who were incarcerated during Anfal, talking about the overcrowded prison halls, the shortage of clean water and food, the spread of epidemics and death, I found it hard to envisage what the women suffered and how they managed.

I became further interested in women's marginalised voices and, in particular, women's experiences of violence and its aftermath while I was doing my PhD on the mental health of Kurdish women refugees. I wanted to know more about the women victims and survivors of Anfal – how they have coped with violence, loss and rupture, what support they have had access to and what role they have played in rebuilding Kurdistan after the No Fly Zone was set up to protect the Kurds in 1991, and a Kurdish parliament was selected for the first time in 1992. It was to answer these questions that I started this research. I was lucky to obtain a two-year post doctoral scholarship from the Leverhulme Trust to go back home and start searching for answers. Later, I managed to secure a one year scholarship from the Kurdistan Regional Government which helped me complete this research. A part of this research will also appear as a chapter in a book about forgotten genocides.[1]

---

1 The Anfal Campaign against the Kurds: Chemical Weapons in the Service of Mass Murder (2011) *Forgotten Genocides Essays on Oblivion, Denial and Memory*. René Lemarchand (ed.) University of Pennsylvania Press.

Acknowledgements

I am grateful to the Leverhulme Trust for a two year scholarship and to the Kurdistan Regional Government for a one year scholarship which made this research possible.

For the duration of this research I was a guest researcher at The Uppsala Programme for Holocaust and Genocide Studies, The Centre for Multiethnic Research. I am grateful to the Centre's hospitality, and for the valuable support and encouraging conversations provided by Satu Grondahl, Kjell Magnusson, Laura Palosuo, Karen Brouneus, Paul Levine, Tomislave Dulic, Jasenka Trtak, Jelena Spasenic, and Ivana Macek.

Thanks to Humbold University and Zentrum Moderner Orient in Berlin for having me as a guest researcher in May-June 2009, during which I had interesting interactions with the research community there. Special thanks to Andrea Fischer-Tahir and Karin Mlodoch for priceless and stimulating debates about Anfal and the survivors.

I am also grateful to various organisations in Kurdistan that helped me during my data collection especially the Women's Education and Media Centre in Suleimanya and Sarqalla, Khatuzeen Centre in Erbil, New Life for Anfal women in Kirkuk, Anfal Centre in Duhok, Anfal Ministry in Erbil, and the Suleimanya Directorate for Anfal and Martyrs Affairs.

Special thanks to various guides and contacts who helped me in this process and some of whom indebted me with their warmth and hospitality while I stayed in their homes: Ikrama Ghaeb, Kawa Mahidi, Arif Qurbany, Jaza Mihammad, Fairooz Taha, and Ali Bandi.

I am grateful to friends and colleagues who read various parts of this book and provided valuable feedback, in particular Jennie Williams, Joost Hiltermann, John Hogan, Lucy Williams, John Hyman, Paul Segal, Laura Palosuo, Joel Hamilton, Goran Baba-Ali, and Tom Godfrey. Special thanks to Rob Cole who, despite our going through a divorce process, came in at the last moment and read the whole of this manuscript to check for coherence and repetition.

And friends who provided last minute crucial support and information, especially Goran Baba-Ali who tirelessly answered my numerous enquiries, Yasser Ashraf, Rebwar Saeed, Adalat Omar, Abdulkareem Haladni, Dlawar Ala'aldeen, and Shireen Drayee. Not to forget Janet and Tim Williams who gave me refuge in their house towards the end of this book and the long walks and good sea-view helped me manage this task better.

Most of all I would like to thank all the women and men who gave me their time and shared with me their experiences despite the difficulties this posed for some. Without their voices and their willingness to share so much of themselves this book would not be possible.

Dedicated to the courageous survivors of Anfal.

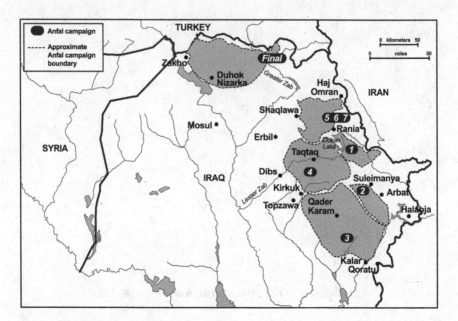

The Anfal Affected Regions

Introduction

Anfal is an under-researched catastrophe and there are limited sources available.[1] Generally speaking, these sources concentrate on establishing 'the truth' about Anfal using Iraqi government documents, forensic evidence, and witness testimonies. Few sources address the problems faced by women during Anfal. Even fewer[2] talk about life after Anfal, namely: the long term consequences of mass violence, the effect of the change in women's social status when they became sole breadwinners in their own families, the destruction of family structure and the farming community, the poverty, the fate of children and the possibility of women's exploitation within their own communities while working as labourers, servants and factory workers. This is true, even though women and their children constitute the majority of the survivors.

This book provides an overview of Kurdish women's experiences and their understandings of what happened to them during the different stages of Anfal and in the aftermath. It addresses the social and psychological consequences of violence, displacement, illness, poverty, and the disappearance of loved ones as experienced by the women. It gives voice to a poor and marginalised group whose stories, and lives, have been exploited by many sections of Kurdish society, including the government, media, researchers, employers, and others. It also investigates the survival strategies of women in the aftermath of genocide. It is hoped that this book will help inform policy and strategies designed to help this group of women, as well as women who have similar experiences in others parts of the world.

After establishing the historical context, and some basic facts about Anfal in Chapter 1, the rest of the book attempts to present the various ways women were affected by the campaign and how they survived. The situation of women who were captured by the army and were detained in prison camps is explored in Chapter 2. These were mainly women who lived in villages inland, away from Iraq's borders with Iran and Turkey, and who were deprived of crucial support that could have secured their escape. The chapter outlines the women's struggle with life in detention as they survived food deprivation, illness, birth, death of children, and the threat of sexual abuse.

The experience of internally displaced women who managed to escape and went into hiding (because of help provided by Kurdish collaborators and relatives,

---

1   There are three main groups of sources, namely the Kurdish sources which include Qurbany (2003a, 2003b), Abdulrahman (1995), Kareem (1996), and Resool (2003); the English language sources including Middle East Watch (1993), Physicians for Human Rights (1993), Makiya (1993) and Power (2002); and articles published on the web.

2   Adalat Omar (2003, 2007).

or because of luck) is explored in the third chapter, along with the experiences of those who escaped to Iran and Turkey. The forced displacement and disempowering situations women found themselves in is outlined here. Even though both internally displaced people and refugees suffer from uncertainty about the future, disempowerment, exposure to hunger, illness, and insecurity, the women who had gone into hiding were particularly vulnerable because of the fear of being reported and being at the mercy of their hosts- relatives and acquaintances that harboured them at this dangerous time.

Chapter 4 addresses the problems faced by women survivors of the gas attacks that were launched between April 1987 and September 1988. Issues regarding ongoing health problems, the fears and vulnerabilities that women experience as a result of uncertainty about the long term consequences of gas attacks, stigmatisation in the community, and lack of financial and medical support are discussed here. This is followed by women's hard work in rebuilding their lives in Chapter 5. After the September General Amnesty in 1988, the surviving women were released from detention camps, those who had gone into hiding came out, and some of the refugees returned to Iraq. This chapter discusses women's hard labour, how they managed the change of status, faced social problems, and dealt with the consequences of poverty for themselves and their children. Women's exploitation in the aftermath and help provided by the Kurdish government are also addressed here.

In Chapter 6 the mental health consequences of violence, rupture, forced displacement, as well as women's survival strategies in the aftermath are discussed. Women's coping skills are discussed in the context of a patriarchal society which disempowers and marginalises them. Anfal surviving women continue to tell their story to each other and to others in an attempt to reclaim their stories and redress marginalisation and exploitation. Finally, the main issues are summarised in Chapter 7. Recommendations are presented for dealing with the problems faced by Anfal survivors, drawing on their own experiences.

The survivor's experiences and recollections are given centre stage in this research in order to showcase their problems, needs, and demands. Women survivors are not mere victims, they are strong survivors. Victimisation of women is addressed alongside their resilience and agency. It is hoped that this book will establish the following issues. First of all, to challenge the dominant Anfal narrative that portrays women survivors as hopeless victims and change the common perspective to see the women as the strong survivors that they are. Secondly, to show how violence ruptures the human bonds essential for identity and attempt to repair these. Thirdly, to address the women's claims for justice, which is a muli-dimensional concept for the survivors. And finally, to discuss the obstacles to mourning the losses caused by Anfal in the hope of achieving closure.

Why Research about Anfal Women

And women are always among the most vulnerable, regardless of circumstances.
(Solomon 2003: 5)

Two categories are of great importance to this research, these are gender and poverty. Being female, a core component of women's identity, means being a member of the oppressed gender group. A Muslim woman from the Middle East grows up in a male-dominant culture. This implies that generally there are more opportunities for men and the culture practices double-standard treatments. These factors may get reflected in a woman's image of the world and of herself. For example, polygamy is still widespread in the Kurdish community even though the Kurdish parliament has imposed increasing restrictions on this practice. Women inherit half of their brothers' share in wealth and property, and two women witnesses together are equal to one male witness in court. The latter was practised in Iraqi Kurdistan until 2008, when the law changed as a result of pressure from the women's movement.

In the more traditional Kurdish regions women are treated as a commodity. They are exchanged in marriages, sometimes without their consent, given away for marriage to settle a tribal conflict or given away for money, commonly known as shirbayee. Women are not allowed to move out of their parental home until they are married. They are expected to satisfy the social demands on them and to put the needs of their families and communities before their own. These norms are commonly internalised by women, even though they restrict their choices and limit their opportunities.

Women are often socially pressurised by family, friends, community, and self to conform to gender stereotypes and this is negatively associated with adjustment and adaptation to new situations. Girls whose lives are shaped by these pressures may never develop 'agentic competencies' which are crucial for facing difficult challenges in life (Egan and Perry 2001). Child rearing practices discourage independence in girls, making them believe that happiness is to be found in selfless service to others (Rampage 1991). Social norms and expectations regarding appropriate female behaviour, rather than biology, discourage women from seeing themselves as capable and independent knowers, able to make choices about their lives regardless of the approval or disapproval of others.

Like other Middle Eastern women, Kurdish women experience low status in their societies. There are restrictions on their behaviour, movements, appearance and clothing (Espin 1996, Elmadmad 1999). Women survivors of Anfal, however, do not suffer merely in terms of their gender; they also suffer in terms of belonging to the poor and uneducated lower class.

According to Weber (1998) gender and class (as well as race, ethnicity, and sexuality) are socially constructed power relationships which empower one group and disempower another. In this sense being 'Male' is privileged and being 'Female' is not, being 'rich' is privileged and being 'poor' is not. Gender and class are deeply embedded in our everyday lives. These concepts are internalised by people, along with ideologically determined explanations for these social relations (Weber 1998: 13). Burstow (1992) refers to this when she says that on top of the damage inherent in oppression, women are damaged by the myths they are fed to cover up the oppression. Women are told that their role as selfless and absolute nurturers who put everyone else's needs first is determined by their nature, as opposed to socialisation. Women, Burstow points out, internalise these myths and they feel guilty if they see themselves failing to satisfy these expectations. This, the author argues, confuses, oppresses and humiliates women.

A gendered approach to Anfal is necessary because of women's susceptibility to violence, including sexual violence, the role of women as the main nurturer and protector of their children, and the balance of power privileging men and undermining women in the Kurdish society which leads to differences in their ability to cope with risks and manage their lives (WHO 2000, Solomon 2003: 5, Williams 2004). This approach is also necessary because, first of all, men and women had different experiences during Anfal. The majority of the men were killed within days (as well as some of the women) and it was the women who lived through the hunger, illness, loss, and desperation in the camps, and they were later released to face the social, economical and political consequences of the campaign. Secondly, men and women, because of gender socialisation, 'experience, remember, and recount' similar events differently (Bos 2003: 33).

While this research was being conceptualised, the questions and concerns were discussed with friends, Kurdish researchers, and writers. Exploring gendered experiences was at the heart of this research. This included issues concerning how women coped in the camps in the absence of change of clothes and sanitation, how they coped with menstruation, pregnancy, birth, illness, death of children and the threat of rape, as well as the situation of women who were forcibly displaced by Anfal and those who went into hiding during the campaign, the experiences and health problems of gas survivors, and issues relating to rebuilding life after Anfal, without support from their patriarchal communities.

Talking in the context of the Holocaust, Bos (2003: 23) points out that 'the history of the Holocaust cannot be told without looking in depth at the experiences and narratives of half of the population which experienced it, that is, at (Jewish) women as subjects'. Similarly, measuring Anfal by death and destruction, and concentration on aggregate mortality figures, property destruction, and the destruction of the Kurdish revolution has produced a gender-blind and therefore an impoverished understanding of Anfal. 'When it comes to death' a man argued, 'how can you worry whether women had clean clothes?'

Generally speaking, women's experiences are regarded as irrelevant to the collective memory of a historical catastrophe. Ringelheim (1997: 20) talks about Pauline, a holocaust survivor, who was molested by relatives of the people who sheltered her during the atrocities. She points out that Pauline kept wondering whether this molestation was important and whether it was part of the Holocaust story. Ringleheim argues that although Pauline recognised that her experiences of the Holocaust were different from men's, she 'did not know how or where to locate them in the history of the Holocaust.' For many years Pauline did not even talk about her experiences to anyone, she was not sure whether they mattered to the Holocaust story. As Ringelheim (1997: 20) argues, this disregard of certain experiences may lead to the memories being ignored by survivors and eventually being 'forgotten.'

During the interviews it was difficult to get women survivors of Anfal to talk about gender specific experiences. A woman who had given birth in Dibs camp[3] failed to mention this while talking about her experiences in prison until a concrete question was asked. It seemed that she considered the specific difficulties she faced as not relevant to this research. Leydesdorff et al. (1996: 5) point out that it is listening to the 'hidden voices' of women and other oppressed groups that makes us question objectivity in history. In other words, if history is written by the dominant group then their version of events prevails over other versions and their voice excludes other voices. This dominance in turn influences what will or will not be remembered:

> The intertwining of power and memory is very subtle … memories supportive
> of the maintenance of existing power structures are usually assured wider social
> space and easier transmission. But memories of subordinate groups can also
> show striking resilience, and they can be transmitted, as women's memories
> often must be, from the interstices of society, from the boundaries between the
> public and the private. (Leydesdorff et al. 1996: 8)

It is true that the majority of the people who were massacred during Anfal were men but staying alive does not mean that women survived unscathed. Throughout her interview, when talking about loss and humiliation, Keejan,[4] who lost her husband and three children, kept wishing that she had died but 'a person doesn't die,' not when she wants to. A woman who hinted at being raped kept saying she wished that she and her children had all died and that 'awful thing' had not happened to her. Another woman who was tortured because of her activities in the camp confessed that she had been a peshmarga. 'What else can you do to us?' she told the interrogator, in the light of the fact that her husband had been taken away, her son starved to death in the camp and she herself was a prisoner. 'There are worse things,' he replied, and by that he did not mean death. While talking about a man who survived the Holocaust when his wife, child and parents did not, Langer (1998: 362) asks an important question: 'Shall we celebrate the fact that because he was a man, and able to work, his life was saved? I think that he, a man crying, would not agree.'

---

3   This woman did not want to be interviewed but she agreed to talk to me about her experiences off the record. Her reason was that she was fed up with giving interviews to people who benefitted from her story and forgot about her afterwards.

4   Keejan, December 2005.

Another common reaction was to argue that speaking of gender specific experiences, in particular sexual abuse, does not help the women survivors. 'Why revive their wounds?' a man insisted, 'If you care about them you must not bring up subjects that would stigmatise them.' This 'moral stance' which attempts to silence all talks of women's private pains is very challenging. Here it is argued that being silent about crimes committed against women is not to women's advantage, that silence amounts to complicity, that as a community we must acknowledge the various ways in which the different groups have suffered in order to heal. That said, addressing these issues has been approached with caution and care. Remembering and recounting sexual abuse, even after 20 years, can still be dangerous.

Das (1997: 84) pointed out that when she was trying to get women to talk about abduction and rape during the partition of India, she found 'a zone of silence around the event.' The language women used to talk about the partition 'was general and metaphoric… evad[ing] specific descriptions of any events.' One woman warned Das that it was dangerous to remember. It is dangerous because women feel that expressing the injustices they have experienced by being victims of rape and assault may only bring them more pain and shame and in some cases social exclusion and death. Thus Das (1997: 85) found that women talked about 'hiding pain, giving it a home just as a child is given a home in the woman's body … this holding of the pain inside must never be allowed to be born.'

It would not be true to say that sexual abuse during Anfal has never been addressed. In February 2003, an Iraqi government document of 1988 was published on a Kurdish website which names 18 Kurdish girls and women from the age of 14 to 29 who were allegedly sold to brothels in Egypt.[5] This naturally flared up the debate about sexual slavery and abuse for a short time. Kurdish nationalists feed on the emotional wounds of Kurdish women being raped and abducted by Arab men. Nonetheless these concerns did not generate research interest into these issues.

Butalia (1997: 93), when addressing women's experiences during the partition of India, states that 'in recovering histories of those who are relegated to the margins, we have little option but to look at sources other than the accepted ones, and in doing so to question, stretch, and expand the notion of what we see as history.' This is because memory is elusive and not without 'contradictions' and 'ambivalences' (Butalia 1997: 94). Similarly Bos (2003: 24) argues that when introducing gender as an analytical tool 'culturally dominant and male ways of categorising what is historically important and what is not are challenged.' Including the sidelined experiences and memories of women in the dominant narrative will only enhance our understanding of Anfal.

There is no intention to reject the 'facts and figures' that make up the Anfal narrative in favour of another version based on women's narratives. It is, however, important to question the accepted narrative and to expand it so that it includes women's experiences. Neither is there an intention to suggest that women are

---

5   Kurdish Media, 2003. Top secret Iraqi document reveals Kurdish girls sent to harems and nightclubs in Egypt. www.*KurdishMedia.com*, 2 July.

absent from the Anfal narrative. On the contrary, women are regularly interviewed by the media and spoken about by the Kurdish politicians. The question is how women are represented in this master narrative. A particularly popular image in the media is that of women lamenting their 'disappeared' husband and children by singing a lullaby.

While talking about Indian women mourning the death of their loved ones Das (1997: 68) stresses that grief is articulated through the body 'by infliction of grievous hurt on oneself' and is 'given a home in language' during lamentation. Similarly, the image of grieved women lamenting the loss of their loved ones, and sometimes inflicting pain on themselves became a typical image of the Anfal women, an image which they were expected to sustain every time they were visited by journalists. These images have been repeatedly used by the Kurdish media to convey the horrors of Anfal. Later they were used by the Kurdistan Regional Government during the election campaigns to urge people not to forget what happened to them under a non-Kurdish government and to encourage people to vote for them (see Chapter 6).

For the women in this research, the aftermath of this catastrophe is as much a part of the Anfal story as the facts and figures that make up the grand narrative. These stories, women's lives, and their complaints about the Kurdistan Regional Government (KRG) are not part of the dominant Anfal narrative. After the establishment of the Safe Haven and the formation of the Kurdistan Regional Government in 1992, the Iraqi government lost control over a large part of Kurdistan. Here an opportunity arose for the Kurdish revolutionaries to address the needs of this aggrieved and violated community and to document and conduct research about Anfal. The KRG failed to make amendments for the Anfal survivors and this has lead to a lot of resentment and pain.[6] It is important to acknowledge the pain of Anfal, and the betrayals and false promises that followed in order to take steps towards healing for the survivors and for the community as a whole.

Methodology and Sample

This book draws on research interviews carried out between 2005 and 2010. The aim was to get a general overview of the women's experiences who suffered during the different stages of Anfal and in the aftermath. The sample included 71 women and 23 men. Various others have contributed to this research through conversations and meetings that took place during fieldwork. Some of the participants were interviewed more than once for different projects. It is interesting to compare the two interviews by each informant and see how they differ (see below). The names

---

6   In the last few years things have gradually improved as the government has provided salary, land and housing. See Chapter 5.

of all survivors have been changed to protect their identity and prevent further stigmatisation in the community.[7]

Various women's organizations were approached,[8] as well as Anfal NGOs,[9] activists,[10] and governmental organizations.[11] Personal and family contacts were also utilised to reach the women in the different regions. The sample included women who had been a few years old during Anfal to older women who were already in their fifties when the campaign took place. The majority of the sample was either illiterate or had had few years of basic education. There were few educated women, who were married to peshmarga, and even fewer who were university graduates.[12] Despite efforts to spend an equal amount of time and give equal attention to all the affected areas this was not entirely successful. Many of the organizations and individuals approached had various contacts in the Garmian region (the third Anfal). Having suffered the largest blow during Anfal Garmian is the place where most organizations have provided a range of projects. Many individual researchers have also focused on this area.[13]

Time was spent in each region, sometimes staying at people's homes, usually people who acted as gatekeepers and guides to the various communities. One limitation with relying on guides was that only survivors who were on good terms with the guide were introduced to this research. To get around this limitation a snowballing technique was used and the women themselves pointed out other women who would be willing to talk. Generally, the women were visited in their homes but sometimes they came to the organization that put me in touch with them. Semi-structured in-depth interviews were used and some of the women (particularly in Garmian) were revisited long after the interviews had been conducted for informal conversations and to see how they lived their lives.

Unfortunately, two of the women interviewees have already died. Saza was killed by her own relatives at the age of 27 in a murder culturally known as 'Honour killing' (see Chapter 5) and Sabiha died at the age of 55 as a result of

---

7   The inhabitants of the following villages and towns have been interviewed for this research: Sergallu, Halladin, Challawa, Kanitu, Sreje, Awaje, Halabja, Deleje, Sewsenan, Aliawa, Masoyee, Bangol, Qochali, Kulajoi Hamajan, Sey Jejni, Durraji, Goma Zard, Takia Kon (Zangana), Qawalli, Sallayee ban shakh, Golama, Mamsha, Sallayee, Qaranaw, Ali Mistafa, Qalla Quta, Khalo Baziani, Wirelle, Koshk, Balleyee Gewre, Dorajee, Tallawa, Warani, Kani Qadir, Znana, Kani Qadir, Mehe Baram, Goptapa, Choghlija, Aspindara, Khora, Warte, Shekh Wasan, Balisan, Ware, Derkari Ajam, Bandaw (Doski), Girgash, Mirgatie, Barchi, Sger (Amedi), Birgini, Warakhal, Razika (Amedi), Koreme, Chelke, and Zinawa.

8   In particular the Women's Education and Media Centre in Suleimanya, Khatuzeen centre in Erbil, Jini nwe bo jinani Anfal (New Life for Anfal women) in Kirkuk.

9   The Anfal centre in Duhok and Erbil.

10   Arif Qurbany, Ali Bandi, Adalat Omar.

11   The Ministry of Human Rights and Anfals, Erbil, 2005 and Suleimanya Directorate for Anfal and Martyrs' Affairs, 2006.

12   For example Nian.

13   Arif Qurbany (in Kurdish) and Karin Mlodoch (in English and German).

chronic illness after being gassed in Balisan in 1987 (see Chapter 4). This brings home the urgency of this research. As the survivors gradually die their voices and stories will be gone with them. It is important, therefore, that their stories are heard and that they play a role in shaping the history of this period.

After years of recounting their stories to each other and to various strangers-journalists, researchers, governmental and NGO workers – survivors have developed long and cultivated narratives. Without stopping or waiting for questions they immediately start talking about their Anfal experiences. It is almost like pressing the play button on a tape recorder. In a similar fashion to other survivors, it was sometimes difficult to ask a question and disrupt this flow of information. Bos (2003: 28) stresses that 'Survivors seek to represent themselves through their narratives in a certain way; they are not likely to concur easily when someone poses a challenge to their particular memories or the depiction of these memories.' Sometimes when a woman was interrupted to ask a clarifying question she looked baffled and even hurt. The best way around this was to let survivors tell their story as they wished, and then later get back to some of the raised issues to get more detail.

Breaking the cultivated narratives down and focusing on certain points proved to be difficult. Sometimes the women mocked the questions. If, for example, they were asked whether there were washing facilities in the camps some women snorted or said, "What are you talking about Miss?" They stated that what they experienced could not be recounted in words. It was not easy to describe, to capture, to bring alive. Despite difficulties, various prompts were used to get the women to focus on certain issues. Specific questions were asked about weather conditions, colour of clothes, smells, particular memories about their disappeared loved ones, specific examples of those who were particularly vulnerable and suffered the most, what they dreamt about, and their hopes for the future.

There is no doubt that the relationship between narrative and truth is a complex one. Hollway and Jefferson (2000: 32) pose important questions when they ask: "What is the relation of a story to the events to which it refers? How is truth compromised by the story teller's motivations and memory?" Bos (2003: 25) asks a similar question when she says: 'How should we interpret the relationship between these narratives and historical experience?' The author then goes on to argue that the fact that 'narratives are generally assumed to be trustworthy historical sources' (p. 29) and 'testimony is read or interpreted as if it were a reflection of an easily accessible truth' (p. 30) is astonishing. This is because, Bos points out, narratives are reconstructions of reality, and experience 'is both uniquely personal and positional, influenced by the different lenses and discourses through which we at different times understand and describe ourselves' (Bos 2003: 30).

In this research there were interesting disparities between different interviews conducted with the same person. This was particularly obvious in the case of two participants. The interviews with a man who was a government official and an older woman showed how testimonies changed over time. Although the bulk of their story was the same, certain events were highlighted and others were dropped.

The tone of the interview was rather different. The elderly woman, two years older and lonelier, gave me more details about people's suffering in Nugra Salman camp and she cried and seemed more vulnerable than the first time I had interviewed her. The man, now a more powerful government official, was more confident and laid back. The details he provided in the second interview were less to do with victimhood and more to do with highlighting himself and his friends as heroes.

Some of these different interviews, on the other hand, were strikingly similar. A woman described her traumatic escape to Iran in February 1988 while she witnessed the freezing to death of many people (see Chapter 3 and 6). These difficult recollections were extremely visual and exact in both of her interviews. This supports scientific findings which suggest that 'when high levels of adrenaline and other stress hormones are circulating, memory traces are deeply imprinted' creating 'sensory and iconic forms of memory' (Herman 1997: 39). Hence traumatic memories are engraved in the mind and they remain as vibrant images, sometimes in a sea of confusing background information.

It is inevitable that narratives are influenced by various factors to do with the witness's state of mind, mood, current position and status. Bos (2003: 31) states that for the witness a selection takes place on three levels: the experiences the witness chooses to talk about, her memory (what she remembers or chooses to remember), and the narrative style (structure, tone, and order). Criticising feminist scholars who over-emphasised certain aspects of women's experiences during the Holocaust (for example, arguing that women, unlike men, were more likely to help each other and therefore were better at surviving), Bos (2003: 36) points out that on the level of experience men and women, conforming to gender socialisation, are more likely to highlight different aspects of their experiences in the camp (women highlight cooperation whereas men highlight independence). This does not mean that women helped each other while men did not. It is therefore important to recognise the role of these different factors and not to rely on them as the truth about what happened in the camps.

Perhaps the best approach is to accept that even though oral accounts deal with how people perceived and experienced the events rather than the events themselves, they are invaluable sources of information. In this research the accounts produced by witnesses, who had lived in the same camps but did not know each other, were largely consistent. As Langer (1991: xv) points out 'Factual errors do occur from time to time, as do simple lapses,' but this seems trivial because the various testimonies built up a complex picture of the events and people's responses. This research is more concerned about how participants perceived the events rather than the events themselves, because it was these perceptions and experiences that informed their choices of action, affected their wellbeing, and shaped their demands for the future.

This book does not claim to capture 'the entire truth' about what happened to all the women who were caught up in the Anfal campaign. To make such a claim would be absurd. We cannot recount the experiences of those who died or provide a census of the vast number who were affected. Neither is there a claim to have provided an in depth picture of the women's experiences. In this effort to capture the various aspects of their experiences, some important issues have not been explored in detail. It is hoped that in the future, researchers will be able to shed more light on various aspects of women's experiences during and after Anfal. Nevertheless, this book by presenting a representative sample provides a platform for the voices which have been for far too long ignored and silenced.

Chapter 1
The Anfal Campaign

I will kill them all with chemical weapons! Who is going to say anything? The international community? Fuck them! – the international community and those who listen to them … I will not attack them with chemicals just one day, but I will continue to attack them with chemicals for fifteen days… Then you will see that all the vehicles of God himself will not suffice to carry them all. (Ali Hassan Al-Majid, known to the Kurds as Chemical Ali, cited in Hiltermann 2007: 95)

The term 'Al-Anfal' means 'the spoils of war.' It is the name of the eighth chapter of the Qura'an which came to the prophet in the wake of his first jihad against non-believers. In 1988 this word was adopted by the Iraqi government for an operation which mainly targeted Kurdish Muslims in the north of Iraq (other, non Muslim minorities such as the Asyrian, Chaldian, and Ezidi Kurdish communities which lived in this region were also victimised). The religiously coined word was used to legitimise this campaign – portraying Kurds as non-Muslims – and to mobilise support for it inside the country and in the Muslim world.

Anfal consisted of a series of eight military offensives that annihilated Kurdish rural life between February and September 1988. The operation brought complete devastation with buildings razed to the ground, water sources blown up or concreted over, and animals, farming machinery and personal belongings looted. It targeted six geographical locations.

This targeted region, home to thousands of farming communities, is where Kurdish resistance to Saddam's dictatorship was most active. It is no coincidence that this region became the principal target of his vicious repression. During the campaign over 2,600 villages were destroyed[1] and an estimated number of 100,000 civilians were murdered.[2] This includes people who were shot in the mass graves and died as a result of the shelling and gas attacks, life in the prison camps, and during their flight to Iran and Turkey. The remaining women, children and the elderly were released during the General Amnesty that September. They were forcibly relocated to housing complexes on the main highways and were left without compensation or support.

---

1 Shorsh Haji (1990).
2 The number given by Kurdish politicians was 182,000. In response to this Ali Hassan Majeed famously said: They were not more than 100,000. Human Rights Watch was able to collect over 50,000 names and estimated the total number to be somewhere between 50,000 and 100,000. Speaking with other field researchers in Kurdistan such as Najmadeen Faqe Abdullah, Arif Qurbany, Adalat Omar and Goran Baba-Ali, I believe the number is closer to 100,000.

Background to the Anfal Campaign

Iraqi Kurds fought for autonomy from the first day of their official forced inclusion into Iraq in 1926. For a short while in the early 1920s, there was hope for an independent Kurdistan as the possibility was left open in the Treaty of Sevres, 1920. This treaty emerged as a result of the Ottoman Empire's disintegration in the wake of the First World War and it was vigorously rejected by the Turkish nationalists. Led by Mustefa Kemal (Ataturk), the Turkish National Movement developed relationships with France putting an end to the Franco-Turkish War (Chaliand 1993: 6). They also defeated the Greek and Armenian forces in 1922 (van Bruinessen 1997: 118). These events and the importance of Turkey as a bulwark against communism forced the former wartime Allies to renegotiate. The Treaty of Lausanne was signed in 1923, invalidating the Treaty of Sevres. Turkey managed to secure the largest part of Kurdistan. By this time, however, southern Kurdistan (Mosul *villayet*) was attached to Iraq which had a British mandate.

The status of southern Kurdistan (northern Iraq) remained in limbo for few years. In 1922, after months of unrest, the Anglo-Iraqi alliance reassured the Kurds that 'His Britannic Majesty's Government and the Government of Iraq recognise the right of the Kurds living within the boundaries of Iraq to set up a Kurdish government' (McDowall 2004: 169). This was abandoned four months later in the Treaty of Lausanne. In December 1925, the League of Nations decided that Mosul should be included in Iraq and the promise of independence was watered down to securing Kurdish cultural, linguistic, and administrative rights by the newly formed Iraqi government (van Bruinessen 1997: 152). Yet the Iraqi government failed to implement these requirements. This led to the break out of unrest in the region in 1930 and to writing a petition to the League of Nations demanding independence. In its memorandum to the permanent Mandates Committee, Britain 'shied away from the inadequacies of its protégé and its own neglect by stressing how wrong headed the Kurdish petitions were in assuming the League had promised them any form of separation' (McDowall 2004: 176).

There were a number of reasons behind the absence of a Kurdish state in this region. The main reason may be the discovery of oil. Britain wanted to seize control over the oil fields in Iraq including the Mosul oilfields (Chaliand 1993: 5). It was decided that there would be one contract with one country and things would be less complicated for the United Kingdom. Another factor that played a role was Turkish insistence on denying the Iraqi Kurds independence. This would 'eliminate the danger of a cross boundary autonomy movement' for the Kurds in Turkey (McDowall 2004: 170). The third reason may be the political immaturity of Kurdish leaders and the failure to present a unified front. The latter was used as an excuse by Britain. However, similar tribal conflicts and lack of unity in the Arab world did not amount to statelessness.

Despite Iraq's record of abuse, it is the only country in the Middle East that legally recognises the existence of the Kurds. This was a condition imposed by the League of Nations for attaching the Mosul *villayet* to Iraq. Nevertheless, negotiations with the central governments over autonomy, its scope, and geographical boundaries continued to be a problem. Successive Iraqi governments tightened their grip in the Kurdish region but the Ba'ath government that came to power in 1968 was the harshest of all. The Ba'ath Party slogan stated 'One Arab nation which has an immortal message' (*Umma Arabia Wahida, thata rilsala khalida*). The insistence that the people of Iraq are part of the united Arab nation automatically excluded the Kurds and portrayed them as an obstacle to this alleged unity, and therefore as a potential threat.

Mala Mustefa Barzani's revolution of the sixties culminated in the 1970 accord with the new Ba'ath government. The 'Autonomous Region' was to include all areas of Iraq where a planned census would determine as having a mainly Kurdish population. This census, however, 'which would have been decisive in the oil rich Kirkuk area, was never carried out' (Chaliand 1993: 7). In 1975 when Iraq struck a deal with Iran in the Algiers Treaty, the Barzani revolution was crushed and a unilateral accord was announced by the Iraqi government. The 'Autonomous Region' excluded the oil rich areas of Kirkuk, Khanaqeen and Sinjar. The government intensified its Arabisation process in these regions. Kurdish civilians were deported to the Arab south or fled to the mountainous region and their homes were given to Arab settlers. After a year of silence, the Kurdish opposition movement resumed its activities in the mountains in 1976.

The eight-year long Iraq-Iran war broke out in 1980. At the time, the Kurdish resistance movement consisted of a number of parties which were in periodical conflict with each other until 1986. The main two parties were the Patriotic Union of Kurdistan (PUK) and Kurdistan Democratic Party (KDP). The smaller parties included the Iraqi Communist Party, the two Socialist Parties, and the Labour Party, all of which had bases in rural Kurdistan. In the 1980s, the Iraqi government gradually lost control over the rural Kurdish region to the peshmarga. This was mainly due to the engagement of the Iraqi army in the war against Iran. The government started its village destruction programme in this region and the villagers were relocated to housing complexes near the main cities.

Between 1980 and 1988, prior to Anfal, 1650 villages were destroyed by the Iraqi government (Resool 2003: 40). The deserted villages were then re-inhabited by the peshmarga. Gradually, some of the deported villagers returned to their lands in the peshmarga controlled region and rebuilt their homes. Some families returned because their sons had deserted the army. Others came back because their land was their only source of income as there was nothing for them in the housing complexes.[3] Throughout this period public services, including teachers, doctors and nurses, were gradually withdrawn from the Kurdish villages.

The villages that were closer to the main cities were under Iraqi control in the day but during the night the peshmarga had the upper hand. Each night when the Kurdish fighters visited a village, the next day the army would raid the village and interrogate its people. Sometimes the men were randomly punished, beaten and

---

3   Nian, March 2006.

even killed.[4] The villagers were crushed between the government and the Kurdish revolutionaries. The majority of them were poor farmers who barely produced enough food for themselves and their families. Yet they were expected to feed the peshmarga. Throughout the 1980s the more active the peshmarga were in a region, the more it was bombarded by the Iraqi government.

In the heat of Iraq's war with Iran, Saddam Hussein tried to make a concession to the Kurds. In the light of PUK's neutrality in the Iraq-Iran war (the KDP had already aligned itself with Iran), the Iraqi government started negotiations with the Patriotic Union of Kurdistan at the end of 1984. Another reason for this negotiation may have been the fact that the PUK was the main political party in the mountainous region along the border of Iran. It had major support in the Soran region, mainly in the cities of Suleymania, Erbil and Kirkuk.

The PUK was initially established by leftist intellectuals in 1976, after the breakdown of the Barzani revolution. Being a younger party with modern ideology, it attracted the intellectual elite as well as many brave fighters. In the 1980s, the PUK was probably the greater threat to the Iraqi government, compared to the other Kurdish parties. The increasing numbers who joined the PUK and the party's continuous activities in the major Kurdish cities required military attention which the Iraqi government could not afford at the time. By engaging in negotiations, the Iraqi government was buying time. It also aimed to isolate the PUK from the rest of the Kurdish resistance movement and to cause more conflict between the factions.

Even though the PUK managed to reach an accord with the Iraqi government, Saddam Hussein refused to announce or implement the results (Resool 2003). Finally, at the beginning of 1986, hostilities resumed between the Iraqi government and the peshmarga. Sanctions were imposed on the Kurdish controlled region and it was subject to continuous bombardment. The Iraqi government vigorously resumed its village destruction programme and the relocation of the villagers to high security camps. In the build up to Anfal, the villages were bombarded on daily basis. Many families did not sleep in their villages anymore because of the unpredictable timing and intensity of the bombardments. Some built shelters and underground rooms near their farms and others slept in valleys and caves.

In the aftermath of the collapsed negotiations, the Kurdish factions started negotiating with each other. In July 1987 agreement was reached and they formed the Iraqi Kurdistan Front (Resool 2003: 61). With the radicalisation of Iraq's military attacks the Kurdish factions had agreed, in what turned out to be a fatal move, to collaborate with the Iranian government. In October 1986, the Iranian army, accompanied by peshmarga forces, attacked the Kirkuk oil fields, infuriating Saddam Hussein (MEW 1993).

It is unclear whether this collaboration with Iran was the main reason behind Anfal or whether it gave the government an excuse 'to solve the Kurdish problem' once and for all (MEW 1993: 58). In March 1987, Ali Hasaan Al-Majeed, known

---

4   Ruqia, November 2005.

to the Kurds as Chemical Ali, was appointed the Secretary General of the Northern Bureau. With this appointment the preparation for Anfal started. Al-Majeed passed a number of decrees and administrative orders such as suspending the legal rights of the villagers, prohibiting food-supplies or contact with the villages, killing wounded civilians and anyone who was caught in the 'prohibited zone' (MEW 1993: 77-80). The villagers were given the same label as the peshmarga, they were called 'saboteurs.'

In April 1987, less than one month after his appointment, Al-Majeed attacked Jafati and Balisan valleys with poison gas and this caused a large number of casualties (see Chapter 4). In the absence of outside intervention and the international community's silence, the planning for Anfal shaped up. The main blow came in the approach to the October 1987 census. The government called on villagers from the prohibited region to register and hence accept relocation to the housing camps built for them. Families who were already living in the housing complexes and cities were subject to removal to the prohibited region if they had a relative, deserters or peshmarga, who failed to 'return to the national ranks' (MEW 1993: 86, see Chapter 2).

Those who failed to register in the 1987 census were no longer considered Iraqi citizens and thus the road to their destruction was paved. It is not clear whether all villagers were aware of the call for registration in the National Census. Even for those who were aware it is not clear that they knew what it meant. The majority of those who learnt about it decided to stay put, unaware of the consequences of their complacency. They believed that by not registering they would lose their Iraqi citizenship and their entitlement to government-issued food rations. By this stage, however, the villages were already without public services and the villagers had already lost their right to food rations. Most people felt that the worst had already happened. Hence the stage for Anfal was set, conceptualised by Al-Majeed himself and carried out under his strict supervision. The villagers became victims in exactly the same way as the peshmarga.

A group that played an important role in the Anfal operation was the jash forces. These were Kurdish tribesmen who were armed by the Iraqi government to fight the peshmarga. The majority of the jash leaders, known as mustashars, guided the army to the remote villages, some of which did not exist on the map, and to hideouts in the caves and mountains. They also deceived the villagers by announcing false amnesties and giving people 'the word of honour' that they would not be harmed.[5] This led to the surrender of many civilians who were never to be seen again. They looted the homes of the villagers and some of them were given medals of bravery by Saddam Hussein. Other mustashars helped people escape at the last minute. It seems that some of them naively thought that the government was relocating the villagers. When they realised what was really going to happen,

---

5   A number of survivors confirmed the same thing.

they quickly switched loyalty. It was due to the help and guidance of members of the jash forces that some civilian lives were saved.[6]

The Circumstances of the Anfal Attacks

Anfal took place in eight stages targeting six identified areas. It went on for six and a half months. At the beginning of each stage chemical attacks were used to kill, terrify and destroy the morale of the people (Hiltermann 2007). After the air attacks and alongside conventional bombing the ground attacks started. The army attacked from several fronts, leaving one way out for the civilians. The attacks were designed to steer civilians towards collection points near main roads where they were awaited by the jash and the army. The civilians were then taken in tractors and military vehicles (IFA trucks) towards the forts and army camps which acted as assembling and processing centres and then they were taken to the temporary holding centres (Topzawa Popular Army Camp during Anfals 1-7, and Qala't Duhok during Anfal 8). They were then divided into three main groups: the men and teenage boys, the women and their children, and the elderly.

The women were transferred to Dibs prison (in the Soran region) and Salamiya near Mosul (in the Badinan region). The elderly were taken to Nugra Salman on the border of Saudi Arabia (in the Soran region) and Duhok fort in Nizarka (in the Badinan region). The men, who were immediately selected for destruction, were stripped down to their sharwal (Kurdish baggy trousers) and vests. Their hands were tied and they were blindfolded and taken to the mass graves. Six men who survived the mass graves were interviewed by Middle East Watch in 1991. They had been taken to their execution squads in the night. They were supposed to be shot by the edge of the empty pits (see Middle East Watch 1993: 239-258). The men's execution was carried out with urgency. This was less so in the case of women and children who were deported throughout the campaign. They were transported from Topzawa and Dibs camps, sometimes after weeks of detention (for more details see Chapter 2).

An account provided by Abdul-Hassan Muhan Murad, one of the bulldozer drivers who took part in digging the mass graves and covering the dead, confirms how caravans of military vehicles were brought full of blindfolded men whose hands were tied. Eleven soldiers did the shooting, according to Murad. Each one of them took out a blindfolded man and brought him to the edge of the grave. The soldiers shot the men and went back to the vehicle to bring out another eleven. Throughout the shooting the bulldozer drivers and the drivers of the IFA trucks and other vehicles were told to keep their engines running to cover up the sound of screams. The women and children were brought out in groups and shot together without having their hands tied, possibly because they were not perceived as a

---

6   Peri, November 2005.

threat (Qurbany 2004). The execution of the families carried on long after all the men had been destroyed.

Anfal 1

The first Anfal attack targeted the Jafati valley where the headquarters of the Patriotic Union of Kurdistan (PUK) was based. It took place between 23 February and 19 March 1988 (MEW 1993: 93, Resool 2003: 87). The valley is located in the mountainous region north of Suleimanya, on the Iranian border. Until February 1988, the harsh terrain had protected the peshmarga from conventional military attacks. This is why the government started the campaign using chemical weapons. The villages that were attacked with gas included Yakhsamar, Sargallu, Bargallu, Haladin, Chokhmakh, Gwezeela, Chalawa and surrounding mountains.

In the heat of the attack the PUK decided to open another front with the Iraqi government to deflect attention away from Jafati valley. In the middle of March PUK forces, along with units from other Kurdish parties, and with support from the Iranian troops, occupied Halabja. This led to a brutal retaliation by the Iraqi government: the gassing of Halabja on 16 March 1988, in which thousands of civilians died immediately. Many others have died since as a result of cancer and other illnesses.[7]

Soon, the peshmarga realised that this time they would not be able to ward off the army. In the midst of the harsh winter, and with mountain routes blocked by snow, they had to quickly think of something to protect the families from destruction. The peshmarga, with help from civilians, cut out an escape route to Iran in the snow. Once the route was cleared they encouraged the families to leave. This was easier for those who were closer to the border, others struggled cutting through the snow. More than 80 people froze to death in Kanitoo region while trying to cross into Iran.[8]

No civilians were arrested by the army during the first Anfal, the majority who survived the snow made it to safety. However, some survivors were arrested in April on the border of Iran where they had taken refuge. Others left the Iranian refugee camps and surrendered to the army lured in by false amnesties in April, May and July. These groups were all arrested and Anfalised (see Chapter 2).

Anfal 2

The second Anfal took place in Qara Dagh region, between 22 March and 2 April. This area is in the southwest of Suleimanya. The villages that were chemically attacked included: Jafaran, Belekjar, Sewsenan, Mesoyee, Serko and Qara Dagh mountain. Although there were orders for holding the displaced villagers in special camps, the rounding up was 'less systematic; compared to later stages of Anfal'

---

7   For a detailed account of the Halabja catastrophe see Hiltermann (2007)
8   Nian, March 2006.

(MEW 1993: 118). The majority of the Qara Dagh inhabitants fled northwards to Suleimanya and the housing complexes. Some of them were arrested on the Qara Dagh-Suleimanya road. Others were arrested in April, during the curfew and house by house search of Suleimanya and the two housing complexes of Naser and Zarayen. The arrested civilians were then taken to the Suleimanya Emergency Forces (Tawari) where the men were separated from the women. The men were never seen again and the women were transported in IFA trucks to Topzawa camp and later to Dibs camp. The rest of the population fled to the Garmian region (warm country) and they were later arrested during the third Anfal (see Chapter 2).

Anfal 3

The third Anfal targeted the Garmian region between 7 and 20 April. This region is in the east of Suleimanya, distant from the Iranian border. The large flat plains of the Garmian region left little chances of hiding or escape. Known as the bloodbath of Anfal, the largest number of women and children disappeared from here. Tazashar was the only village bombarded by chemical weapons. The flatness of the region and the dense army presence ruled out extensive gassing. The army attacked from several fronts, steering civilians towards Qader Karam. Some of the villagers bribed the jash or were helped by them and secretly sneaked into Laylan and Shorsh near Chamchamal. Others were arrested in their own village as their homes were looted and bulldozed before their eyes. They were then trucked away to their different destinations.

To lure people in the jash announced a false amnesty during the third Anfal. The promise was broadcast by mosque loudspeaker in Qader Karam (10-12 April). Some of the *mustashars* gave their word of honour that no one would be harmed. The large civilian disappearance and in particular the loss of women and children on 14 April made this date of particular significance.[9] It remains a mystery why such a large number of women and children disappeared during the third Anfal. One reason may be the closeness of this region to the Arabised oil rich city of Kirkuk. The mass disappearance may be another method used by the government to change the demography of the region once and for all. This is supported by Staff Major-Gen. Wafiq Al-Samara'i, deputy of Iraq's military intelligence director in 1988, who stated:

> You can kill half a million Kurds in Erbil, but it won't change anything: It would still be Kurdish. But killing 50,000 Kurds in Kirkuk will finish the Kurdish cause forever. (Al-Samara'i, quoted in Hiltermann 2007: 134)

---

9 Najmadeen Faqe Abdullah established the first organisation to collect data about Anfal and it was due to the research of this group that 14 April became recognised as the national commemoration day for Anfal.

Middle East Watch (1993: 170) found that in the areas where peshmarga resistance was fiercest, the largest number of civilians disappeared. It is also possible that as the attacks went on, and the government realised that the international community was not going to intervene, the policies became harsher. This is true of the last Anfal in Badinan where extensive chemical attacks caused great casualties. Another reason may be that Garmian consists of flat plains and is totally landlocked, away from the border. In other words the government's policy may have been opportunistic.

Anfal 4

The fourth Anfal was in the Valley of the Lesser Zab, in the region of Kirkuk and Koysinjaq. It took place between 3 and 8 May. Peshmarga from the third Anfal retreated to Askar and Shwan in the region that became prey to the fourth Anfal. The fourth Anfal started with major gas attacks on Askar and Goptapa. The people panicked and started heading in different directions. Another false amnesty was announced and hundreds of men came out from their hideouts and surrendered (Middle East Watch 1993: 178). In some villages the men were separated from women at point of capture while their homes were being looted. Others were separated in the forts and camps.

Anfal 5, 6, and 7

The fifth, sixth and seventh Anfal targeted the valleys of Shaqlawa and Rawandiz in the Erbil district, on the border of Iran. This consisted of three consecutive offensives which started on 15 May and ended on 26 August. The villages which were gassed include Ware, Seran, Balisan, Hiran, Smaquli, Malakan, Shekh Wasan, Rashki Baneshan, Kaniba, and Nazaneen. The fifth Anfal attack failed. Two more attacks were required to bring the region under government control, hence the 5th, 6th and 7th Anfal offensives. The peshmarga commander of this region, Kosrat Abudllah, prepared for a long siege by stocking up on food and ammunition. He also evacuated many families from the region and made a deal with the jash leaders to allow people to escape.[10] Some of the villagers had already fled after extensive chemical attacks in April 1987 which caused a large number of casualties in Balisan, Shekh Wasan and Malakan. The remaining civilians fled in different directions. Some fled to the border of Iran to the east, some headed to housing complexes near Lake Dukan in the south, others hid in mountains and caves.

Anfal 8

The final Anfal targeted the Badinan region which was a KDP stronghold. It took place between 25 August and 6 September. Chemical attacks were used against

---

10   Kosrat Abdullah, Suleimanya, November 2006.

Zewa Shkan, Tuka, Glenaska, Birjini, Tilakru, Barkavreh, Wirmeli, Bilejane, and Ikmala. The panicked population started migrating towards Turkey. On 26 August Iraqi troops blocked the escape route to Turkey. Those who managed to cross the main road or were closer to the border managed to escape, the rest were arrested and transported to the holding camp. The men were separated for execution while the women were detained in Fort Duhok and then in Salamya near Mosul until the September Amnesty.

The September General Amnesty

> The men all disappeared, half of the women also disappeared and they released
> the rest for propaganda. They took the young and released the old and useless.
> (Runak, December 2005)

By September 1988 the Iraqi government had achieved a large part of its aims. The Kurdish countryside was destroyed and all men aged between 15 and 50 had been killed or forced to flee. The Kurdish resistance movement was crushed and its remaining supporters had fled to Iran. The whole of Iraqi Kurdistan was brought under control by terror and the desired stillness and silence was achieved. This, however, was not the only reason why Iraq released the surviving women, children, and elderly.

In August 1988, when tens of thousands of Kurds who were fleeing the final Anfal sought refuge in Turkey, reports were leaked by journalists about extensive use of chemical weapons after the war with Iran had come to an end. These reports alarmed Saddam Hussein about potential confrontations with the international community. He was clearly worried about being charged with genocide long after the campaign had come to an end.

Peter Galbraith of the US Senate Foreign Relations Committee, who had visited the Kurdish region in 1987 and seen the mujama'ts (housing complexes) that were made for the Kurds, concluded that what was happening in Iraq was 'genocide'.[11] He was the main force behind The Prevention of Genocide Act 1988 which was unanimously passed by the US senate even though unknown to the West, the genocide had already come to an end. Galbraith's work started after Anfal ended, mostly because Anfal remained well hidden from the public view outside Iraq (Hiltermann 2007). This Act, however, was rejected by Reagan's administration because it believed that this was 'too strong a response.'

In September 1988 both Physicians for Human Rights and the US Senate Foreign Relations Committee sent teams to investigate the alleged use of chemical weapons. Both of them concluded that these weapons had been used. The Genocide Act, although not adopted by the US government, was a strong enough message

---

11   Peter Galbraith, 16th March 2007. Anfal: The attempted destruction of the Iraqi Kurds seminar. Centre for Study of Holocaust and Religious Minorities, Olso.

to the Iraqi government. This led to the first large-scale anti-US demonstrations in Baghdad and Saddam Hussein did not use gas against the Kurds again. Not even in 1991 when, in the aftermath of the uprising, nearly two million Kurds fled to the borders of Iran and Turkey in fear of these weapons.

In the aftermath the Iraqi intelligence services were keeping an eye on all activities which might bring attention to Anfal. On 24 June 1989, the Istikhbarat (military intelligence) Northern Sector issued a document concerning a Press Conference held in Stockholm which was addressing Anfal and where pleas for international interference on behalf of the Kurds were made.[12] In 2003, a Kurdish artist, Pishko Abbas, was interviewed on Kurd Sat TV. The artist lived in Italy during the Anfal campaign. In 1989 he put together a solo exhibition about Anfal. He was approached by some men who spoke fluent Italian and claimed to be journalists. They requested to interview him at home and take some pictures. On the day of the supposed interview, the minute he opened the door to them he realised they were not Italian journalists, but Iraqi agents. He tried to remain calm and to escape through his kitchen window when they captured him. He was arrested, handcuffed, taken to the Iraqi embassy, put on a plane, and then taken back to Baghdad. In Baghdad, he was interrogated and tortured, put on trial and given a death sentence because of being 'a traitor.' He was in solitary confinement waiting to be hanged when an amnesty was issued for political prisoners and he was released. Immediately after his release he escaped the country once again and went into hiding for a time.

The surveillance the Iraqi government carried out on people and the attention they paid to any mention of Anfal inside the country and abroad shows that Saddam Hussein was concerned about being accused of genocide. Releasing the remaining civilians in the General Amnesty may have been to deflect such accusations by showing that Anfal was a counter insurgency measure and that civilians were not killed but merely deported. However, probably half of the village population had already been destroyed, largely the men and teenage boys but also some women and children.

Another explanation may be inherent in the nature of the Ba'ath state's policies of terror.[13] During its rule the Ba'ath government carried out various raids and curfews during which people were arrested, interrogated, tortured, and convicted of conspiracy and sabotage. At the same time the government issued amnesties to political prisoners, some of whom were released days before execution. The release of tortured, traumatised, and broken individuals into the society was another form of asserting control through terror. The purpose was to let the rest of the community see what could happen to them if they were to defy the government. In a similar fashion, releasing the traumatised Anfal prisoners into the community, and the stories that circulated about their ordeal, may have been an attempt to pacify the rest of the Kurdish community into submission.

---

12   Iraq Documentation Project. Page 1273980.
13   This idea developed as a result of a converstion with Goran Baba-Ali.

Anfal and the Question of Genocide

We went to Anfal in tribes, we came back as individuals. (Gelawej, December 2006)

The trial of Saddam Hussein, and seven high ranking Iraqi officials, for the crime of Anfal started on August, 21, 2006. Initially, Saddam Hussein and Ali Hassan Al-Majeed were charged with genocide, war crimes and crimes against humanity. However the execution of Saddam Hussein at the end of December 2006, before the Anfal trial was concluded meant that the case against him was dropped. Al-Majeed faced the charge of genocide alone. The five other defendants were charged with war crimes and crimes against humanity. The defence argued that the operations were a legitimate counter-insurgency measure, targeting pro Iranian Kurdish guerrillas. Sultan Hashim Ahmad who was commander of the third Anfal operation said that civilians were not targeted, in fact they were 'safely removed.'[14]

The defendants denied the use of gas attacks. To counter this denial the Kurdish television channels repeatedly played a tape recorded meeting of the Northern Bureau and its Ba'ath directors in the northern governorates, in which Al-Majeed talks about 'the deportation.' He stresses 'I will not attack them with chemicals just one day, but I will continue to attack them with chemicals for fifteen days' (MEW 1993: 349). During the trial Al-Majeed first expressed doubts about the authenticity of the tape and then went on to say 'If I had said such things it was merely to frighten people into submission but we never used gas attacks.'[15] The Chief Prosecutor pointed out that this information was disclosed in a secret meeting between high ranking military and party officials and the information was not for public dissemination in order to frighten civilians. Al-Majeed also claimed that he had 'never before heard of Nugra Salman and Dibs camps', neither had he heard of 'mass graves.'[16]

According to Article 2 of the UN Convention for Genocide, 'genocide means any of the following acts committed with intent to destroy, in whole or in part, a national, ethnical, racial or religious group, as such: (a) Killing members of the group; (b) Causing serious bodily or mental harm to members of the group; (c) Deliberately inflicting on the group conditions of life calculated to bring about its physical destruction in whole or in part; (d) Imposing measures intended to prevent births within the group; (e) Forcibly transferring children of the group to another group.'[17] To establish whether or not genocide was committed by Iraq will depend on establishing 'intent' in 'acts committed with intent to destroy.'

Helen Fein (1993: 10) stresses that 'intent' is 'purposeful action.' I would argue that 'intent to destroy the group' can be inferred from the conception, meticulous

---

14   BBC (2006).
15   The Anfal Trial, broadcast on Kurdistan TV. 7 February 2007.
16   The 40th session of the Anfal trial. Kurdistan TV.
17   Convention on the Prevention and Punishment of the Crime of Genocide, Article 2.

planning, execution, and success of the campaign as shown by the captured Iraqi documents, the mass graves and the testimony of survivors. It is, therefore, necessary to establish that the killing of civilians (by gassing, shelling, and shooting in the mass graves), causing serious physical or mental harm (gassing, shelling, arresting, detaining, beating, destroying a way of life, and killing members of the family), and imposing life conditions that brought about death (through starvation diet, lack of sanitation and the spread of epidemics in the camps) were deliberate actions to destroy the group as opposed to accidental consequences of another action – a counter-insurgency campaign against the peshmarga.

The targeted group in this campaign is a sub-group within the Kurdish community in Iraq, namely the mountain dwellers. In this sense, 'a part' of this ethnic group was targeted for annihilation: the Kurds who lived in the villages. This 'part' was selected for annihilation in two stages: first by declaring the areas 'prohibited for security reasons' (Resool 2003: 76, MEW 1993: 82) and then by imposing the condition of registering in the 1987 census. All those who lived in the prohibited region and who failed to register were automatically selected for destruction.

Like other cases of genocide, other ethnic groups were also targeted within this campaign.[18] Officially there were two recognised groups in Iraq. These were Arabs and Kurds. Other ethnic groups such as the Turkomans, Assyrians, Chaldeans, and even Ezidi Kurds were officially registered as Arabs. The fact that villagers from these communities did not register in the October 1987 National Census meant that they had denounced their Arab identity and therefore they were perceived as siding with the Kurds and they too were targeted. In fact their punishment was more severe because the September General Amnesty excluded them (MEW 1993).

According to Fein (1993: 34), there are three similarities between groups of victims that become prey to genocide. These are: firstly, 'they are alien or perceived as alien' and hence are 'outside the universe of obligation of the dominant group.' Secondly, they are seen as 'un-assimilable' or may themselves refuse to assimilate, and finally, their destruction either 'gets rid of a threat or opens up opportunities.' The Anfal victims satisfy all three qualities. First of all, all Kurdish villages were pronounced 'prohibited for security reasons.' Failure to register in the 1987 National Census meant that they were not considered Iraqi citizens anymore and therefore they were 'outside the universe of obligation.' Secondly, the Kurds in general have rejected assimilation in Iraq, but this may be particularly true of rural Kurds who refused deportation to the housing complexes and chose to remain in their ancestral home. Even though this meant living with continuous bombardment and threats. Thirdly, the destruction of Kurdish villages would mean that the Kurdish opposition movement would lose its support and access to food and shelter. In this sense, the destruction of Kurdish villagers meant the destruction of the Kurdish opposition movement and was therefore getting rid of a threat.

---

18   The Holocaust included Roma, Poles, political prisoners from various countries and the Armenian genocide included Asyrians and Pontic Greeks.

Following Fein's model, Anfal was retributive genocide. The Iraqi government was retaliating for the Kurdish opposition's attempts to challenge 'the structure of domination' (Fein 1993: 87). Since Iraq's establishment the Sunni Arabs, who are a minority in Iraq, had been the dominant ruling class. The Kurds have repeatedly revolted against this domination – Sheik Mahmud's revolt in the 1920s, Barzani's resistance in the 1940s and later in the 1960s, and then from 1976 till 1988 Talabani, the Barzanis and others. Hence, the history of Iraq provides one precondition for genocide.

Kuper (1981: 57) points out that 'plural ... societies with persistent and pervasive cleavages' between different racial, ethnic or religious groups offer a necessary condition for 'domestic genocide.' Kuper (1981: 59) also suggests that 'Colonisation is a major creator of plural societies.' This is true, not just in the case of countries that were colonised by the industrialist Western powers, but also when these same Western powers made decisions about how to divide up regions into nation states. They drew lines across the map, dividing certain communities and forcibly putting others together. The way the borders were drawn in the post First World War Middle East meant that the Kurds, who consider themselves one community, were divided. They were made part of nation states in which they constituted a minority and where they were dominated and oppressed. This is despite the fact that they were a large community with viable economic resources, and the will to establish a nation state.

Kuper (1981: 59) then goes on to stress that in many plural societies, with a history of ethnic or religious conflict, genocide does not arise. This may be for 'lacking the subjective reactions and opportunity to sustain a destructive conflict.' The rise of the Ba'ath Party provided the subjective reaction necessary for genocide. The party's Arab nationalist ideology excluded the Kurds and portrayed them as a potential threat. At the beginning, however, the young Ba'ath government did not have the power or capacity to fight the Kurds and thus tried to negotiate with Barzani (1970-1975) and then with Talabani during the war with Iran (1984). Later, as the Ba'ath government's monopoly on power was established, the war against the Kurds intensified through Arabisation of the oil rich areas, destruction of Kurdish villages, and repression through torture, imprisonment and disappearance.

In response to the Iraqi state's repression the Kurdish opposition also became radicalised and anti-government activities were carried out across Kurdistan. The opportunity then arose to 'solve the Kurdish problem' during the Iran-Iraq war and with the acquisition of weapons of mass destruction. The Iraqi government was at first incapable of destroying Kurdish opposition, partly because of the type of guerrilla warfare which was fought by the Kurds from tough mountain peaks that seemed insurmountable. The chemical weapons destroyed the protection provided by mountains and caves. These weapons were successful in casting terror amongst the Kurds. They broke the civilians' spirit and confidence. The ancient Kurdish saying, 'No friend but the mountains' was no longer true, even the mountains were not friends anymore.

The radicalisation of Iraq's policies towards the Kurds culminated in genocide. At first all public services were withdrawn from the villages and they were routinely bombarded. Then they were declared out of bounds and all exchanges between the cities on the one hand, and the villages on the other, were forbidden. This was followed by the request to register in the National Census, and finally it ended in destruction. For genocide to take place there needs to be an organised perpetrator who defines the victim group, collects its members and destroys the group. Fein (1993: 25) states: 'Documenting genocide requires at least identifying a perpetrator(s), the target group attached as a collectivity, assessing its numbers and victims, and recognising a pattern of repeated actions from which we infer the intent of purposeful action to eliminate them.' She then goes on to identify five propositions which are necessary and sufficient conditions for genocide. I will introduce each of these propositions and show how Anfal satisfies all of them.

1. The perpetrator launched continuous attacks to destroy the group (Fein 1993: 25). In the lead up to Anfal the military was ordered to become more efficient in killing civilians. Clause 4 of a personal directive by Al-Majeed (June, 20: 1987) ordered the military commanders to bomb these regions with 'artillery, helicopters and aircraft at all times of the day or night in order to kill the largest number of persons present in those prohibiting zones,' maximising the number of victims.

The operation itself took place over six and a half months (23 February to 6 September). Six regions were identified, attacked from various directions until all the inhabitants were killed or arrested (some managed to flee) and all the houses and water works were destroyed. The methods of killing used were direct by bombardments, gas attacks, and shooting at the mass graves (clause (a) of the UN convention), and indirect by exposing the group to harsh conditions of life in the prison camps which brought about the death of many (clause c). Food deprivation, spread of disease, and lack of sanitation and clean water resulted in the loss of many lives. Soon people, and in particular children, were dying of starvation and disease. The latter continued in the housing complexes even after the Amnesty of September 1988. Most families had lost their breadwinners, their means of earning (farming), and their entitlement to food coupons and government support. Those who survived in the aftermath only managed because of the support and charity of locals.

In April 2005, a mass grave containing the bodies of 1,500 civilians was discovered in the south of Iraq, the majority of the victims were women and children.[19] Gunned down in their bright Kurdish clothing, some of the victims are said to have been holding pots and pans and toys, an indication that they believed they would be relocated. Between 2003 and 2006 six mass grave sites were exhumed while preparing for the Anfal trial.[20] Some of these provided important clues as to who the victims were. The forensic team which was led by Michael

---

19   Knickmeyer (2005).
20   Daragahi (2006).

Trimble, found 'an unexpected wealth of identification cards in mass graves.'[21] Many of the women had managed to hold on to their Iraqi Identity Cards, despite being asked to hand them over and despite being searched. Through these cards the victims were traced back to where they came from, and surviving relatives were found to testify against Saddam Hussein.

2. The perpetrator was leading an organised group of actors (Fein 1993: 25).   From the time of his appointment as Secretary General of the Northern Bureau in March 1987, Ali Hassan Al-Majeed issued decrees and held meetings with high ranking army and security officials to tell them about the plan. These meetings were all tape recorded and many of them, along with Al-Majeed's personal directives and decrees, were captured by the Kurds in 1991. His 'decisions and directives' were to be obeyed by all agencies including intelligence, security, and Popular Army command (Middle East Watch 1993: 57). He also checked on the different regions to make sure that his orders had been carried out to the word. There was continuity of leadership throughout the campaign under Al-Majeed. The military, security and intelligence agencies were all organised and ordered by him.

It is estimated that the government deployed 50,000 troops, including Kurdish mercenaries against the 2000 peshmarga who defended the Jafati valley in the first Anfal (Resool 2003: 88) and as many as 200,000 troops against 2,600 peshmarga in the final Anfal of the Badinan region (MEW 1993: 264). The Iraqi Air Force including fighter aircraft and helicopters were involved, as well as tanks and truck mounted multiple barrel artillery (rajima). The main section of the army which carried out the ground attacks involved infantry, Popular Army (al-jaysh al-sha'bi), Commandoes (mughawir), National Defence Battalions (jash), Special Units (quat khasa), The Republican Guards (haras jumhuri), the Security Units (amn) and Intelligence Officers (mukhabarat) (Resool 2003: 88).

3. The victims were chosen because they were members of the identified group (Fein 1993: 25).   The victims were selected merely on the basis of being Kurdish (or Kurdish-identifying) inhabitants of villages. They were all pre-selected before Anfal, through a series of tightening decrees issued by Al-Majeed from 1987 till 1988. On 20 June 1987 all Kurdish villages were pronounced 'prohibited' (Resool 2003: 76, MEW 1993: 82). Signed by Al-Majeed himself, the directive was addressed to the military, the security directorate, Istikhbarat and Mukhabarat. This was the first stage of defining and pre-selecting the chosen group. From this date, all the Kurds who lived in the 'prohibited region' were called 'saboteurs' and treated the same way as the peshmarga. More than 2,600 Kurdish villages were thus pronounced 'prohibited', and the murder of inhabitants was obligatory.

The second stage of selection started in the run up to the National Census of 17 October 1987. Throughout seminars and meetings, Al-Majeed stressed that 'any persons who fail to participate in the census without a valid excuse shall lose their

---

21   Associated Press (2006).

Iraqi citizenship' (MEA 1993: 87). Thus the inhabitants of the 'prohibited zone' had two choices. Either they would register in the census, and be relocated to Iraqi government controlled housing complexes (mujama't) or fail to register and lose their Iraqi citizenship. Stripping people of their citizenship was the final stage of selection for destruction. Those who failed to register were to be segregated and killed, even if they surrendered.

4. The victims were killed regardless of whether they surrendered or resisted (Fein 1993: 25). The majority of the people who disappeared during Anfal were villagers. Most of these villages had been repeatedly attacked by the government throughout the 1980s. As the war between Iraq and Iran continued and the bombing intensified, the villagers learnt to live with the war, attending to their crops in the night. Even the inhabitants of the villages that had been destroyed earlier, and had accepted deportation to the mujama't were rounded up, they were expelled to the 'prohibited zone' just before Anfal if they were related to a 'saboteur.'

From April to September, various false amnesties were announced to encourage people to surrender. The jash promised that anyone who surrendered would be saved. Thousands of people who surrendered to the jash were then arrested and disappeared. Four of the mass grave survivors interviewed by Middle East Watch were men who had surrendered to the jash during this period. Similarly, many families from the first and second Anfal who had arrived at safety in Iran, decided to return to the border of Iraq in April. Dozens of families were arrested in Sune, the women and the elderly were taken to Dibs and Nugra Salman prison camps while the men were never to be seen again.[22]

In July 1988, rumours of another amnesty spread amongst the inhabitants of a refugee camp in Seriaz, Iran. A woman and her ten children who had been injured in the Halabja gas attacks joined dozens of refugee families who decided to go home. The families had young children, many suffered from gas burns or illness and wanted to escape the harsh life in the tents. They were guided by some Kurdish men who were later found out to be members of the jash forces. The returnees were awaited by military trucks on the border that took them to Topzawa. In a similar fashion, the men were separated from the women in Topzawa and the women were taken to Nugra Salman camp in the south.[23]

5. The destruction of group members was deliberate and murder was authorised by the perpetrator (Fein 1993: 25). Clause 2 of the personal directive on 20 June 1987 points out that the prohibited zone is 'strictly out of bounds to all persons and animals,' and the troops are given free rein to open fire 'without any restrictions.' In Clause 3, travel to these zones, animal farming and any industrial activities were pronounced prohibited in this region. Clause 4 orders the commanders to bomb the region 'at all times of the day or night in order to kill the largest number of persons

---

22 Qumri, April 2006.
23 Naji, April 2006.

present in those prohibiting zones, keeping us informed of the results.' In Clause 5 Al-Majeed stated that in the prohibited region 'all those between the ages of 15 to 70 shall be executed after any useful information has been obtained from them.' (Middle East Watch 1993: 80). In December 1987, as a result of repeated enquiries by high ranking Iraqi officials demanding clarification of this clause, Al-Majeed repeated: 'The security agencies should not trouble us with queries about Clause 5, the wording is self-explanatory and requires no higher authority.' (Middle East Watch 1993: 81). This is direct authorisation for the killing of victims.

In the run up to the October 1987 National census, Al-Majeed announced that anyone who fails to participate in the national census 'shall lose their Iraqi citizenship. They shall also be regarded as army deserters and as such subject to the terms of Revolutionary Command Council decree no. 677 of August 26, 1987.' This decree stated that 'the death sentence shall be carried out by party organisations, after due verification, on any deserters who are arrested' (Middle East Watch 1993: 87). This order was internally distributed to all sections of army and government machinery, the general public, however, remained unaware of it.

The day after the National Census, the Secretary of Revolutionary Command Council's Northern Affairs Committee, Taher Tawfiq: 'issued a stern memorandum to all security committee in Kurdistan, reminding them that the aerial inspection would ensure the Directive of June 20 was being carried out "to the letter." Any committee that failed to comply would "bear full responsibility before the Comrade Bureau Chief" – that is to say, Ali Hassan Al-Majeed' (Middle East Watch 1993: 89).

Finally, there have been a number of acknowledgements, even though few and far between, of Anfal as a case of genocide. In 1993 Human Rights Watch published a comprehensive report about Anfal, recognising it as genocide. In 1995 the US state department's legal adviser determined that Anfal was genocide (Hiltermann 2002). On 23 December 2005 the Dutch trial convicted Van Anraat, a 64 year old businessman who sold tons of raw materials for chemical weapons to Iraq, of crimes against humanity.[24] The trial ruled that the gassing of the Kurds between 1987 and 1988 were acts of genocide.[25] More recently on 24 June 2007 the Iraqi High Tribunal convicted Ali-Hassan Majeed (and others) for genocide, crimes against humanity and war crimes. This was the first time that Iraqi government documents, as well as forensic evidence from mass graves, and witness testimonies were presented in a trial which was broadcast throughout Iraq. However, criticisms of the trial by international human rights organisations[26] undermined the international recognition of Anfal once again.

---

24 The Associated Press (2007).
25 BBC (2005).
26 Reterska and Sissons (2007).

Anfal Coverage: 1988 and Beyond

Throughout 1988 there were repeated outcries from Kurdish leaders, the international press and human rights organisations about the extensive use of chemical weapons and the mass exodus of Kurdish refugees into Iran and Turkey. A number of gruesome attacks received considerable media attention before Anfal's atrocities emerged in full light. The first major attack was the gassing of Halabja on 16 March 1988. Samantha Power (2002: 187) calls this 'a Kurdish Hiroshima.' She compares the scenes of death and devastation revealed in the wake of the attacks on Halabja to a 'modern version of Pompei.' with the victims 'frozen in time ... some slumped a few yards behind a baby carriage, caught permanently holding the hand of a loved one or shielding a child from the poisoned air, or calmly collapsed behind a car steering wheel.' Not all victims died instantly, she writes, 'some of those who had inhaled the chemicals continued to stumble around town, blinded by the gas, giggling uncontrollably, or, because their nerves were malfunctioning, buckling at the knees.'

Although this attack was launched during the first Anfal offensive (23 February-19 March), it was not part of the Anfal operation. Halabja was a town with a population of about 80,000 and it was not part of the region which the Iraqi government had declared 'prohibited for security reasons.' It was gassed in retaliation for its occupation by Kurdish fighters, assisted by Iranian soldiers, two days earlier. If Halabja did not go unnoticed in the media this is because it was accessible to foreign journalists, being only fifteen miles from the Iranian border. Proximity to Iran also meant that the Iranian television crews were able to film victims. These images were repeatedly shown by Iranian news media in their coverage of the war, and they also appeared in Western media directly after the attack.

The Reagan administration bears a major responsibility for confounding Iraq's role in the gassing of Halabja by stating that 'Iran, as well as Iraq, used chemical weapons in the Halabja incident.' (George Shultz, quoted in Hiltermann 2007: 127). This position was taken not for want of information about the extent of the tragedy but because the United States was Saddam's closest ally – politically, economically and militarily – in his war against Iran.[27]

The second major crime that received media coverage was the final Anfal attack, from 25 August to 6 September 1988, which took place after the ceasefire of August 20 between Iraq and Iran. More than a dozen villages were gassed during this attack, leading to a mass exodus of Kurds to Turkey. Unlike Iran, which was only too keen to expose Saddam Hussein's atrocities, Turkey acted as a complicit bystander. On 3 September, during the last Anfal offensive, the British government spoke of 'grave concerns' about possible use of chemical weapons.

---

27 For details of how the UN, US, Iraq and Iran responded in the aftermath of Halabja see Hiltermann (2007: 125-129).

However, it 'did not favour its own investigation but asked for information from Turkey' (McDowall 2004: 362).

On 9 September 1988, the Turkish Foreign Ministry announced that it had 'no evidence' that the Kurdish refugees were suffering from chemical weapons injuries (Physicians for Human Rights 1989: 4). Following this, in a visit to Southeast Turkey (11-17 September) Peter Galbraith of the US Senate Foreign Relations Committee said he had found 'overwhelming evidence'[28] that Iraq had used chemical weapons.

On 12 September, the United States along with twelve other countries asked the United Nations Secretary General to investigate the matter but Iraq and Turkey denied access to the UN team (PHR 1989: v). The Turkish Foreign Ministry representative asserted that Turkish medical experts had 'found no trace' of chemical weapons. In this light, he argued, a UN investigation was unnecessary because it would 'create a wrong impression that Turkish medical experts are inadequate to make related research.' (PHR 1989: 4)

In October 1988, a team of three doctors representing Physicians for Human Rights (PHR) went to Turkey to investigate the allegations. They relied on meeting survivors in refugee camps who had fled northern Iraq in August 1988. The team was only able to visit two small camps in Mardin and Diyarbakir as they were denied access to the larger camps in Silopi and Yuksekova by regional Turkish officials. They were also unable to talk to Turkish physicians who had initially treated the refugees and were the first to be in contact with them.

In spite of the difficulties they faced, the PHR team found 'convincing evidence' that chemical attacks were used against the Kurds (PHR 1989: vi). However, Iraq continuously denied these allegations and declared this an internal affair. The United Nation's 'protective stance in relation to the sovereign rights of the territorial state' (Kuper 1981: 164) ruled out taking action against a state that engages in genocide against its own people.

Despite these reports, and the controversies that followed, Anfal's nature remained unknown until the Popular Uprising in March 1991. Iraq's defeat in the first Gulf War and America's encouragement of the people of Iraq to topple Saddam Hussein's government led to the short lived popular uprisings in the Shiite south and the Kurdish north. According to Makyia (1993b: 59), the uprising in the south was launched by defeated Iraqi soldiers who returned from Kuwait disillusioned, angry and full of shame. In Kurdistan it started spontaneously by the public who believed that the Americans would support them after Bush's statement to that effect.

---

28   Chemical Weapons use in Kurdistan: Iraq's final offensive. A staff Report to the Committee on Foreign Relations, United States Senate, October 1988.

In February 1988, the PUK radio repeatedly sent messages to the *hallos* (hawks) to get organised and prepare for 'the wedding day.' The old hymns were being replayed: 'No one should say the Kurds are dead, the Kurds are thriving. We are thriving and our flag is never lowered.' Kurdish leaders secretly negotiated with the jash forces and promised them an amnesty in return for their cooperation. It was clear that victory was only possible if the jash did not fight them. Agreement was reached and the peshmarga started sneaking back into Iraqi Kurdistan in March 1991. In the space of a few days most of the Kurdish towns and cities fell to the Kurds.

During the uprising the Iraqi prisons, security and intelligence offices were raided. According to an agreement between the PUK, the Human Rights Watch and the US Senate Foreign Relations Committee 14 tons of Iraqi government documents, which were captured during the raids, were transferred to the US National Archives for 'research and analysis' (Middle East Watch 1993: xxvi). These documents revealed the truth about the government's atrocities in Kurdistan, including the Anfal campaign.

In 1993, a year after the documents were transferred to the US, an important first account was produced by the Middle East branch of Human Rights Watch. This book is a full report of the Iraqi government's conception, planning and execution of genocide (Middle East Watch 1993). In the same year Physicians for Human Rights and Middle East Watch produced a joint report about the destruction of Koreme (MEW and PHR 1993). This village is in the Badinan region, on the border of Turkey. Thirty-two men from Koreme were executed after failing to escape to Turkey during the last Anfal offensive in August 1988.[29]

Anfal and Silence

Despite the media coverage at the time, the reports and documentaries that followed, and the large body of Iraqi documents, witness testimonies, and forensic evidence, Anfal remained largely unrecognised by the international community. When addressing the use of chemical weapons by Iraq, McDowall argues that in 1988 no country wished to take a lead in investigating these claims or producing the evidence they had because:

---

29   Other reports in English include Makiya (1994). The author went back to the Kurdish region in 1991. His journey in Iraqi Kurdistan was captured in a prize winning documentary by the British journalist, Gwynne Roberts. Makiya visited the fort of Qoratu and discovered trucks full of women's clothes and belongings which had been abandoned and covered in three years of dust and rain. He also interviewed Taymour, the twelve year old survivor of a mass grave, about his journey and escape. In this documentary there is a terrifying scene of the Anfal widows beating their chests, tearing at their hair and holding up fingers to show the number of people they have lost during Anfal.

Behind the expressed concern of governments not to jeopardise the Iran-Iraq
peace talks by condemnation of Iraq, lay the real concern not to jeopardise the
massive post war reconstruction projects (estimated at $50,000 million) that Iraq
was bound to put out to tender. (McDowall 2004: 362)

Business and trade links with Iraq may have made the West turn a blind eye to
Saddam Hussein's violations of the UN conventions. But later, and especially
after the first Gulf War when such links were terminated, it was possible to
investigate these crimes. Iraq is a signatory to the 1951 Genocide Convention and
has acknowledged the legality of verdicts passed by the International Court of
Justice. A question remains as to why Saddam was not indicted for crimes against
humanity and genocide. Past relationships with Iraq may have contributed to this
failure.

In 2002, after pressure from the US to disclose his chemical weapons
programme Saddam Hussein submitted a report to the UN. The report disclosed
information about the extensive participation of foreign companies in Iraq's
weapons programme. The US, in agreement with other Western countries, decided
to prevent the release of this information 'in order for these not to fall into the
wrong hands.'[30] According to Zumach (2002) more than 80 German companies
were involved in supplying and supporting Iraq's chemical and biological
weapons programme. The US supply comes second with 'approximately two
dozen companies' as well as providing 'non-nuclear building parts.'

The involvement of American companies in supplying Saddam Hussein with
raw chemical material is of particular significance because senior politicians in the
government were involved in some of these companies. This may be one of the
main reasons why Saddam Hussein was not indicted for Anfal in the 1990s despite
the enormous body of information and evidence. Another reason may be that
during the Iraq-Iran war, the United States provided intelligence on the location of
Iranian troops which also helped the Iraqi government trace and target the Kurdish
fighters in the region.[31]

Most Western countries were unwilling to jeopardise their relationship with
those Arab and Islamic countries that indirectly supported, or at least sympathised
with, Saddam Hussein. During the early 1990s, when Human Rights Watch was
compiling a dossier against Iraq they found no Western country that was willing
to bring a case of genocide against Saddam in the Hague (Hiltermann 2007: xii).
It was also believed that, after imposing sanctions on Iraq, Saddam Hussein was
contained and therefore rendered harmless. Some argued that Saddam's removal
from power might disturb the balance of power in the region. For a long time,
the Ba'ath government was perceived as a bulwark against Islamic extremism
in nearby Iran. Ironically, this allowed Saddam Hussein to remain in power long

---

30   Tageszeitung (2002).
31   Peter Galbraith, 16th March 2007. Anfal: The attempted destruction of the Iraqi
Kurds seminar. Centre for Study of Holocaust and Religious Minorities, Olso.

enough to strengthen the ethnic and religious rivalries which existed from the day of the country's establishment. The current turmoil in Iraq shows us that in the long run Saddam's stay in power may have led to the rise of Islamic extremism and increased destabilisation.

Another factor which contributed to this may have been the absence of an organised and united Kurdish government that could carry out research about the issues and build comprehensive national archives. Over the years Anfal has become a focal point for Kurdish victimhood in Iraq, the principal symbol of their long suffering and all the injustices they experienced. Anfal further strengthened Kurdish demands for independence. It is the perpetual backdrop to most Kurdish political dialogue. In Iraqi Kurdistan the term is now being used as a verb. Those who disappeared in this campaign are said to have been 'Anfaled.' The term has entered common usage and has been stretched to apply to other instances of mass disappearance.[32] On occasions, the application of the word 'Anfal' to other instances of violence is badly chosen. The word is being used erroneously as a synonym for genocide and mass violence.

Conclusion

Anfal was the climax of the Iraqi state's atrocities against the Kurds. It is a logical conclusion to the state's progressive hostility towards its largest ethnic minority. The Kurds, comprising 20 per cent of the population, never welcomed their inclusion in Iraq. They waged many revolts against the various central governments. This requested resources which the government resented. Before consolidating its power in Iraq the Ba'ath government, which came to power in 1968, engaged in a negotiation with the Kurdistan Democratic Party (1970-1975). Later this was repeated in the middle of the Iraq-Iran war when the government negotiated with the Patriotic Union of Kurdistan (1984-1985). Both negotiations fell apart because the Ba'ath government did not abide by its promises to the Kurds. These periods were more like a temporary ceasefire during which the Iraqi state reorganised its forces, preparing for another period of war.

During Iraq's war with Iran (1980-1988) the Kurdish liberation movement was increasingly perceived as a nuisance because of needing military attention that the government could not afford. From early on the KDP, which was based

---

32   In 1983, special units of the Iraqi army arrested between 5,000 and 8,000 males over the age of 12 from the Barzani tribe. The incident took place in Gushtapa, a camp to which the Barzani tribe had been relocated in 1981 (Makiya: 1992). The Barzani clan was particularly targeted because they were a key faction behind the Kurdish liberation movement since the 1960s. It was also a retaliation for the Kurdistan Democratic Party's (KDP) collaboration with Iran during the Iraq-Iran war. The victims of this attack, some of whom were only recovered and returned to the Barzan region in early 2006, are now being called Anfal victims.

in Iran since its defeat in 1975, joined forces with Iran. The PUK and the other smaller parties, however, initially remained neutral. Between 1980 and 1988 Iraq deported 1650 villages near the main roads, creating a safety belt between itself and the peshmarga-controlled region. The mountainous region was then declared prohibited, sanctions were imposed, all forms of communication and movement between the region and the rest of Kurdistan were forbidden, and it was subject to daily bombardment.

After the breakdown of the PUK-Iraq negotiations, and faced with Iraqi policy's radicalisation, the various factions of the Kurdish movement unilaterally decided to form close links with Iran in 1986. This enraged the Ba'ath government. In retaliation the government decided to destroy the whole population in the 'prohibited' region. The rationale behind this decision was that, the government believed, in order to permanently dislodge the Kurdish movement it was necessary to get rid of its supporting foundation, namely the Kurdish village population which carried on producing food and goods oblivious to the state's sanctions, restrictions, and bombardments.

The opportunity arose to get rid of 'the Kurdish threat' through the acquisition of chemical weapons. Iraq tested the power of its weapon on the civilian population in April 1987. In the absence of the international community's reaction the government shaped up its plans. Iraq's abundant use of chemical weapons in this region was mainly to kill but also to terrify and uproot the civilians. The siege of each region then followed ensuring the capture of civilians and their deportation to camps. In the camps civilians were processed into three groups- the men and teenage boys were immediately selected for destruction, the women and their children to Dibs and Salamya camps, and the elderly to Nugra Salman and Nizarka. Many women and children were also murdered throughout the campaign. Convoys of trucks deported the families from Topzawa and Dibs camps. Many of the elderly died because of the inhumane conditions in the camps.

Anfal was not a counter-insurgency measure to get rid of the peshmarga but it was designed to destroy everyone in the region, regardless of age and gender, and whether or not they surrendered. Despite available evidence and news coverage at the time, the world remained untouched by Iraq's atrocities. This silence was the result of a complex set of factors including the involvement of western companies in providing raw material to Iraq's chemical weapons programme, mutual interest and business and trade links, and the unwillingness of Western governments to jeopardise their relationships with other Arab and Muslim countries. As a result Anfal remained largely forgotten and unrecognised.

After 2003, and the collapse of the Ba'ath government, an opportunity arose for Anfal to be recognised within Iraq and in the world. The Anfal trial which was held by the new Iraqi court was severely criticised by human rights organisations and the international community for its shortcomings. Its legitimacy was questioned and Anfal lost out on recognition for the second time. The execution of Saddam Hussein before the trial was over was another major source of disappointment for Kurds. Many felt angry that he was executed for the murder of a few dozen people before he was tried for the murder of tens of thousands. Anfal remains a meaningless name for most Westerners and this continues to cause frustration and embitterment amongst the Kurds.

Chapter 2
Women in Detention

I can talk about Anfal till the next day but you will never understand because you weren't there ... It cannot be written, it cannot be described. There is no end to talking about it. (Razaw, May 2010)

Anfal started in the depth of a harsh winter and carried on through the scorching summer. Driven out of their villages by bombardment, gassing and military siege civilians left their homes on foot, by mule, or in carts pulled by their own tractors. The carts were stuffed full of children, the elderly, women and basic belongings such as food, blankets, and clothing. The army claimed civilians at the nearest main road. Those who were on foot were packed into open backed IFA trucks, while the others were led in their own tractor-pulled carts, to the nearest assembly point. Upon arrival, all their belongings were confiscated and they were left in the clothes they were wearing to face the traumas of Anfal.

From the point of capture onward civilians were processed via a number of camps and forts in which they spent a few days each. The various transfers and long journeys, some lasting up to 14 hours, without access to food, water or toilets, were designed to disorientate and exhaust civilians and break their spirit. The journey in military vehicles, ambulance-like vehicles, and large buses resembles that of Holocaust victims who were crammed into cattle cars, driven for days, and were reduced to going to the toilet in a bucket in the middle of an overcrowded wagon (Vago 1998: 274).

The majority of women were arrested during the different stages as the attacks were proceeding. Others were captured during the curfews, and house to house searches that followed each stage of Anfal. These were women who had managed to escape because of help provided by members of the jash forces, through paying bribes, or by escaping via brief breaches on the main roads. The rest of the women were captured when they surrendered during the various false amnesties. Those who had managed to escape to Iran during the first Anfal returned to the border of Iraq in late spring and early summer. They were surrounded by the army and taken away. A minority of women who had survived the gassing of Halabja made it to Iran and then returned to Iraq. They too were arrested and they got tangled up in the Anfal campaign.

The purpose of detention is to exert control over and isolate individuals (Herman 1997). The result of detention is feeling powerless and, in extreme cases, vegetation (Herman 1997: 85, Matussek et al. 1975: 15). Garwood (2002: 356), a child survivor of the Holocaust and a psychoanalyst who has worked with survivors, argues that the essential components of camp survivors' trauma are fear of annihilation, powerlessness, loss, and torture. He then goes on to say that 'Being forced to be totally passive and helpless in the face of the Holocaust was perhaps the most devastating experience for the survivor' (Garwood 2003: 358).

For the informants of this research, powerlessness, loss, and fear for one's life dominated the camp experience. The victims went through progressive powerlessness. They were forcibly displaced by gassing and bombing, then they were arrested. All their belongings were confiscated. They were separated from family members, and were exposed to filth, hunger, and violence. Fear for one's life was instilled by a starvation diet, lack of sanitation, the spread of lice, bugs and disease, random beatings and punishments (especially in Topzawa, Nizarka and Nugra Salman camps), the decline of other individuals in the camp, and the deportation of those who were taken away in closed military trucks, not allowed to take their clothes or babies' bottles (from Topzawa, Nizarka and Dibs). They were never to be seen again.

While in detention, people's access to food, water, sanitation, and even toilets were strictly controlled by their captors. Nugra Salman, the elderly people's camp, was a particularly bad place to be because of the excessive desert heat, starvation diet, regular beatings, and intense fear of Lieutenant Hajjaj and the black dog that ate the dead bodies (see below). Herman (1997: 77) points out that, in order to gain total control, the perpetrator subjects the victim to 'systematic, repetitive infliction of psychological trauma' which cause fear and vulnerability. This is achieved through the use of actual and/or threat of violence, and the enforcement of 'petty rules.' Also, through controlling the victim's access to food, sleep, washing facilities and clothes the perpetrator destroys the victim's independence and individuality keeping her powerless, preoccupied with basic survival needs, without a choice and subjugated. Similarly Garwood (2002: 357) argues that such 'carefully planned and developed programmes' are designed to weaken, degrade and enslave inmates. This makes them falsely believe that if they obey, they may be allowed to live.

In this chapter, I will present women's experiences in the prison camps. This is divided into three parts. I will first outline the women's journeys, which ended in their imprisonment. In the second part I will describe the physical conditions in the prison camps. The accounts provided by women regarding the general physical conditions during their flight and in the camps were largely consistent with each other. They have, therefore, been presented as they are. In the last section, the more controversial and sensitive experiences of camp life are explored. Issues such as sexual abuse, giving birth, losing children, the factors that helped women survive, and generally what is considered socially and ethically sensitive, have been addressed here. The experiences and accounts have been compared and contrasted with each other, contextualised by women's similar experiences in other parts of the world.

The Journeys

> It was a rain storm. From the land Saddam was oppressing us and from the sky God.
> (Gullalla, December 2005)

In February 1988, when the first Anfal started, the majority of the inhabitants of Jafati valley lived away from their villages because of the regular bombardments. Some had built shelters by the mountainsides and others lived in the caves and valleys. Some civilians, however, remained in their homes and the gassing of 25 February 1988 claimed lives in all the bombarded villages. Haladin was one of the villages which was gassed on this day. At its full capacity the village consisted of 150 families and it had a hospital, a primary school, and a mosque (Resool, 1990: 111). There were relatively few casualties in Haladin considering the size of the gas attack – five people according to some while others provided a larger estimate.[1] The gas attack led to mass chaos and fear amongst civilians who took to the mountains. It was a harsh winter and the mountain routes were blocked by snow.

At the end of February, the PUK forces, helped by civilians and with Iranian assistance, broke into the snow and opened a few escape routes to Iran.[2] 56-year-old Qumri,[3] along with her family and other villagers, escaped via a narrow passage through the snow, leaving behind all her possessions. She kept falling and slipping. The group walked for hours carrying their children, food supplies, and some of their belongings with them. Soon, Qumri recounts, people started abandoning blankets, weapons, radios, and other possessions they had hoped to hold on to. After a few hours some couples considered leaving some of their young children behind because they could not carry them anymore.

In Galala, on the border, this group with other fleeing refugees were picked up by Iranian cars. They were transported first to Bana, then to Sawen and later to Sardasht. Qumri's family stayed in Iranian Kurdish people's homes and in tents in refugee camps. They were cold and desperate. They lived on minimal food. After hearing about an amnesty in April, a group of people decided to get back to the Iraqi border. Qumri followed the crowd with one of her sons, her daughter-in-law and her grandson. The group was surrounded in Sune, arrested and transported in military trucks first to a military base in Erbil, where Qumri was reunited with her older brother, and then to Topzawa army camp.

Qumri's son, his wife and child were separated from her in Topzawa and that was the last time she saw any of them. She was then taken to Nugra Salman camp, where her blind sister had already been deported. Alongside her brother and sister, Qumri was detained in this camp until the September General Amnesty. Similarly,

---

1 The names provided were: Yaseen Abulrahman Aziz and his daughter Srwa Yassen Abdulrahman, Saeed Abdulrahman Salih, and Twana Ali Kareem.

2 Secret cable no. 4610, February 25th 1988, cited in MEA (1993: 98).

3 Qumri, April 2006.

Hajar[4] from Sargallu village[5] made it to Iran through the snow with her three sons, daughter-in-law and two grandchildren. Hearing about the amnesty in April, this woman's family, along with a few others from the region, came back to the border of Iraq. Unlike Qumri's group, who were awaited by the army, this group were not immediately picked up. They had a dialogue with a jash leader who persuaded the men, 17 of them, to surrender. These men were never seen again. The women and children, however, were allowed to sneak back into town and they escaped detention.

While the first Anfal was continuing, Halabja was gassed on 16 March 1988. Naji[6] and her ten children were injured during this attack. That afternoon, they joined hundreds of others who were fleeing to the Iranian border. They were all blinded by the gas and people guided them through the corpses that littered the streets. They stumbled and fell while holding on to each other to avoid getting separated. By dawn the family arrived at Sazan Bridge, on the border of Iran. They had just managed to cross over when the planes came once again and gassed the valley. More people were injured here. After laying low for a while, Iranian cars came to pick up the wounded refugees and took them to Iran.

Four months later rumours of an amnesty spread in Seriaz refugee camp where many Halabja survivors lived. Naji's children jumped at this opportunity and wanted to go back to Halabja. She tried to dissuade them but they were exhausted and they would not listen to her. The family joined the large group, about 1900 people according to Naji, who were planning to go back. They felt safe amongst such a crowd. They were guided by men who were later found to be members of the jash forces. The IFA trucks were waiting for them at the border. When she saw the soldiers Naji whispered to the people around her that they 'have been sold.' The soldiers looked 'unfriendly and grudging' and they acted suspiciously towards the returnees. Naji thought that either they knew what would befall the people and 'felt sorry' for them for coming back of their own will, or 'they hated the Kurds and wanted [them] all dead.'

The returnees, most of whom were Halabja survivors, became victims of Anfal. They were taken to the Emergency Forces (Tawari) in Suleimanya where their names were recorded. After three days, the men were separated from the women and convoys of military vehicles came to take them to their different destinations. While the prisoners were being loaded into the trucks people of the nearby neighbourhoods in Suleimanya were 'crying and beating their breasts' for the prisoners and throwing food into the trucks. The soldiers started beating the women, forcing them back into their homes. Naji was terrified by this sight. She became even more certain that they would not be taken somewhere good. They were transported to Nugra Salman camp on the border of Saudi Arabia.

---

4 Hajar, April 2006.

5 Sargalu consisted of 400 families and it had a mosque, a primary school and a hospital (Resool 1991: 111).

6 Naji, March 2006.

During the second Anfal some peshmarga discouraged people from leaving, hoping that the presence of civilians would help them survive. The inhabitants of Aliawa, a small village which consisted of 10 families,[7] remained in their village despite the battle getting closer. When Sewsenan village was gassed on 22 March killing more than 80 civilians,[8] the people in the region started panicking. Shirin,[9] Habiba[10] and Shara,[11] three young mothers who were related by marriage, were amongst the group who left Qara Dagh and headed to Garmian. Between them they had eight young children and were accompanied by one of their husbands and his elderly mother. It was raining and they had three mules to share. The families walked to Qopi Mountain in the rain and entered Garmian from Tapa Garuz.

Shirin was carrying her one month old baby as she struggled walking through the rain and thick mud. She realised that her son was tightly 'clinging to [her] collar.' Shirin cried as she remembered the baby's fists just as she had cried all those years ago. She kissed the baby and whispered to him not to worry because 'mummy would not leave [him] behind.' During the exhausting journey, which took several days, the family took detours to avoid the main roads. They arrived at Garmian, and right then the third Anfal started. The army was closing in on people: 'A helicopter was circling ahead and one behind us,' recalled Habiba, 'There was no way we could go anywhere else.' On 14 April this family, along with thousands of others from Garmian and Qara Dagh were captured on the road to Milla Sura.

Thousands of women and children were arrested during the third Anfal. Shadan,[12] who is originally from the city, was evicted from Kirkuk four months before Anfal started. The family were deported to the 'prohibited region' because Shadan's brothers-in-law were already living there as deserters. They went to the village of Qochali – a small village comprised of 20 families,[13] where Shadan's husband had relatives. When the third Anfal started the men all hid in the mountains while a jash leader took the families to Soran hoping to protect them without avail. They then came back to the main road to Qader Karam where they were herded together. Dozens of buses and IFA trucks came to take them away on the next day. No one knew where they would be taken.

In an attempt to calm the women down, the mustashars (jash leaders) told them they would be taken to Kirkuk 'for some questioning.' One of the bus drivers, however, knew that they would be taken to prison. He urged the women to escape when they arrived at Chamchamal. 'This is my bus,' he told them, 'shred it to pieces and escape whenever you find the opportunity.' Shadan then burst into tears. She had three young children and knew that she could not run as fast as the others.

---

7 Resool (1990: 149)
8 Bakr, Sewsenan Martyrs' Graveyard, April 2006.
9 Shirin, April 2006.
10 Habiba, April 2006.
11 Shara, April 2006.
12 Shadan, March 2006.
13 Resool (1990: 29)

Her eldest child was three, the other was a year and a half and the third one was five months old. When they arrived in Chamchamal, however, it was impossible to escape. The people of Chamchamal had risen against the government the day before and they had managed to rescue some prisoners. Because of this the region was crammed full of the military. The driver wept and told the women that he could not help them. Shadan remembered how the civilians were 'shaking their heads' from desperation but 'no one had the power to help, no one.'[14] The women were then taken to Topzawa and later to the women's camp in Dibs.

Some civilians were encouraged by the jash to surrender during the third Anfal. As the attack got closer most of the inhabitants of Bangol, which consisted of 38 extended families,[15] decided to stay put.[16] Some people had gone into the hills and orchards, but everyone knew that there was no way out. The mustashars came to Bangol and persuaded people to surrender on the Qochali road. They ate dinner with the people and promised to give the men false jash papers for protection. They advised the men to surrender with weapons in order to qualify for the deal. This, however, ensured their destruction, they were all considered 'sabouters' and 'collaborators' and they were all killed.

One of the young men who did not have a weapon and had to buy one was Keejan's[17] brother-in-law who was a young shepherd. Keejan remembered how naïve the villagers were in those days, how easily they were deceived by the jash forces: 'We believed them because they were Kurds, but they sold us to the government.' The civilians walked to Qochali drenched in rain and mud, carrying their children on their backs. It was only when they met the army did they realise what was going on. The men were aggressively called apart and the soldiers shot down two mules which carried food and children. The children were 'drenched in blood' and no one dared to go and pick them up. At this moment Keejan realised what was going to happen: 'We were going to be annihilated.'

The fourth Anfal started on 3 May with the gassing of Goptapa and Askar villages. Being the larger village of the two,[18] Goptapa sustained the largest number of casualties. Between 160 to 200 civilians died in these attacks.[19] 105 of them are now gathered and buried in the 'Martyr's Graveyard' at the top of a green hill in Goptapa. After the gassing, the inhabitants of the valley scattered in different directions (MEW 1993: 175). Some people fled southwards in the direction of Suleimanya and Chamchamal, while others fled westward. In a similar pattern to the third Anfal many families were arrested from the valley of Lesser

---

14   Shadan, March 2006.
15   Resool (1990: 29)
16   Keejan, December 2005.
17   Keejan, December 2005.
18   Goptapa consisted of 452 extended families and it had a mosque, a primary school and a hospital (Resool 1990: 96).
19   Muhamad, February 2006.

Zab. According to Ibrahim[20] about 400 villagers from Askar and Goptapa fled to Gopala and were arrested there on 4 May. They were taken to Suse complex and then to Topzawa. Only the elderly were ever seen again.

Haji Ali[21] recounted that a dozen extended families from the villages of Goptapa, Jalamord, Tutaqal and Chuqlija, headed towards the Khalkhalan mountain and hid in the valley for seven days. On the morning of 10 May the army surrounded them. On the spot they separated the men and took them away. The elderly and the rest of the families were transported in IFA trucks first to Taqtaq and then to Topzawa. Three days later the elderly were put into one closed vehicle and the families into the others. Haji Ali, who was watching from the small window in the back of the vehicle, kept an eye on the other trucks that carried, amongst others, his daughter with her nine children, his daughter-in-law with her five children and his second daughter-in-law with her new born baby. At a roundabout near Samawa, the convoy divided, the elderly were taken to Nugra Salman and the women and children were taken elsewhere: 'The families we were separated from we never saw again.'

The valleys of Shaqlawa and Rawanduz, where the PUK's third Regional Command (Melbend) was based, held out for longer than three months. This region was attacked three times before it was brought under control, hence Anfal 5, 6, and 7. On 15 May Anfal came to this region with the gassing of Ware a day before the end of Ramadan (MEW 1993: 196). This was replicated on 23 May when Balisan, Malakan, Shekh Wasan, Nazaneen, Seran and Bleh were gassed (MEW 1993: 196). The attack on Ware was devastating considering the size of the village which consisted of 40 extended families.[22] Layla,[23] who was 12 years old, was at home when the bombs were dropped. She inhaled the gas and became hysterical. Her aunt grabbed her arm and dragged her out of the house. As they ran through the village, they saw tens of people falling and dying. The civilians were advised to light fires or immerse themselves in water in case of gas attacks. Since there was no time to light a fire they ran to the water spring, unaware that a gas bomb had dropped into it. People threw themselves in the polluted water and died in convulsions. According to Layla around 40 people died in the village on that day. This is consistent with Middle East Watch findings (1993: 196) that 37 people died as a result of the Ware gassing.

Layla fainted when she got to the other side of the village. She was later picked up by her maternal uncle who lived in Hartal. When she woke up amongst her relatives she was blind, aching and confused. She could not understand what had happened even though she had seen the gassing and the corpses. For a long time Layla acted strangely, and people believed that she has gone mad as a result of the

---

20   Ibrahim, February 2006.
21   Haji Ali, February 2006.
22   Resool (1990: 127).
23   Layla, April 2006.

gas. This woman, however, was not arrested. She was helped by members of the jash forces and her family and she managed to escape prison.

A large part of the population in this region had already left as a result of the April 1987 gas attack on Balisan valley. The first four Anfals were also a good warning signal to the remaining people in the valley. The families were encouraged to either leave to Iran or to sneak back into the housing complexes nearby. Their journey to Iran was facilitated by the peshmarga.[24] Some families were even protected by powerful jash leaders (MEW 1993: 198). Only a small number of families were arrested from the Shaqlawa valley. In some cases only the men were arrested and the women were allowed to go free.[25]

The final Anfal started on 25 August, targeting the Badinan region where the Kurdistan Democratic Party was based. It started with gassing more than a dozen villages (MEW 1993: 270). 16 year old Saadia[26] from the village of Zinava[27] was nine months pregnant with her first baby when Anfal reached her village. She had been married for eleven months. On the morning of 25 August she and her husband were woken by neighbours who told them that they must flee. That morning blinded and wounded villagers from Birgini were seen in the fields 'stumbling and screaming from pain.' The inhabitants of Zinawa walked for three days hoping to make it to Turkey. All the villages they came across were deserted. They saw the gassed villagers and livestock in the fields and slept in the mountains to avoid being caught. But soon they were surrounded and taken away.

Adiba and Fairooz[28] pointed out that their men were hoping to fight alongside the peshmarga, but soon they realised that they could not fight against chemical weapons and decided to flee with their families. About 200 people from Darkari Ajam, 57 men and their families, fled towards Turkey but the army arrested them on the way. They made the people walk to the main road where the men and women were loaded into separate trucks. They were taken first to the military centre (Firqa) in Zakho and then to the Duhok fort at Nizarka, where the women witnessed the stripping and beating of their men. The families, along with hundreds of other families from the region, were kept in Nizarka for up to five days while the men were taken away in closed vehicles. The women were then transported to Salamya in Mosul where they remained until the General Amnesty was announced two weeks later.

The Moment of Separation from Men

Women spoke about the moment of being separated from their husbands, brothers, fathers, sons, and male relatives with great pain. They felt hopeless at how the

---

24   Kosrat Resul, PUK Office, Sueliemanya, November 2006.
25   Hajar, March 2006.
26   Saadia, March 2006.
27   The village consisted of 40 extended families (Resool 1990: 187).
28   Adiba and Fairooz, February 2006.

men were treated at the point of capture. They talked, in detail, about the men being humiliated, beaten by cable, slapped and kicked, shoved around, made to run around, and stripped to their vests and sharwals. Seeing the men powerless, frightened, and oppressed made them feel depressed and hopeless. This resonates with the experience of women survivors of Holocaust who also felt great pain because 'the accepted image of men as strong and independent was far removed from the sight of the helpless and degraded male prisoners they met in the camps. What they saw depressed them. Perhaps seeing this transformation moved them towards a fuller realisation that the world they knew was gone' (Tec 2003: 127).

The men were either forcefully taken away or in some cases just quietly disappeared without any notice. Semeera's[29] husband left her on the road to Kifri, where hundreds of families had gathered, saying that he just wanted 'to check out what is going on over there, in that crowd.' That was the last time Semeera saw her husband. She waited for him for a few hours until his cousin came and told her, 'there are soldiers in that crowd, he's been arrested.' It was so quick that she did not understand that this was the end of him. His disappearance happened in a very casual and quiet manner and there was no time for dramatic reactions, no time to say goodbye. This sudden rupture baffled Semeera for years. She found it difficult to believe that she would never see her husband again.

Gullalla,[30] who last saw her husband in Topzawa, reported that 'till they put the stone on [her] mouth (in the grave) [she] will never forget the pain of seeing [her] husband like that.' As he waited his turn to be taken away in the large overcrowded courtyard her husband, who was about to cry, told her that he was 'finished' and he urged her to look after the children. A few days later the men were brought out of their separate halls, 'kicked into trucks' and taken away. All that was left of them were piles of their clothes, belongings, and ID cards.

Nizarka fort, which was the equivalent of Topzawa in the Badinan region, was another place where women hopelessly watched their men suffer. Saadia[31] went into labour in the truck that brought her to Nizarak. Her father was beaten for helping her off the truck. When she was trying to stand straight in the courtyard she saw the guards setting fire to her husband's beard which in the 1980s was associated with harbouring pro-Iranian sympathies. She wept in pain and fear but did not dare to make a noise. A woman from Saadia's village whose husband had died a few days earlier lost her mind when her two sons were beaten to death in Nizarka. She tore her clothes and started beating herself. She ran to her sons and put herself between them and the blows. She was beaten to death along with them. Later, Saadia's father put her clothes around her and helped bury the woman.

After two failed attempts to escape to Turkey, the inhabitants of Koreme, along with some people from Chelke, were arrested and taken back to their village.

---

29 Semeera, December 2005.
30 Gullalla, December 2005.
31 Saadia, February 2006.

Nalia[32] was separated from her husband in Koreme. Thirty-two men and teenage boys from the villages of Kurme and Chelke were lined up and escorted towards the almond trees to the east of the village. The women and children were steered in the other direction, towards the trucks which would take them to prison. The soldiers reassured the women and told them: 'Don't worry, your men will follow.' Nalia kept turning back and looking at her husband until she could not see him anymore. A few minutes later, endless rounds of shooting made the children jump and everyone started screaming and crying. The soldiers told them it was the sound of friendly fire and there was nothing to worry about. The families had to keep walking towards the trucks but they knew that the men had been killed. Nalia remembered how a soldier hit one of the women who was crying too loudly and told her to be quiet. The families were then loaded into the trucks and taken to Mangeshk school and then to Nizarka. It would be three years before Nalia would find her husband's body in one of the four Koreme mass graves and be certain (see Chapter 6).

The Camps

During the Holocaust there were various 'assembly and transit camps' where people were collected before deportation; 'concentration camps' where people were detained, tortured and forced into slave labour in factories, coal mines, and on the roads (until they dropped dead or were selected for extermination by gassing); and 'extermination camps' where millions were killed by gassing (Ofer and Weitzman 1998: 267). Some of these camps 'contained "laboratories" in which "scientific" experiments were conducted' on human beings (Tec 2003: 122). Those who survived the experiments were disfigured and disabled and most of them ended in the gas chambers (Lengyel 1993). The majority of the people who were incarcerated perished by gassing, but tens of thousands of others were shot or they died because of the overcrowded and inhumane conditions, violence, hunger, and epidemics. There were around ten thousand camps and 5,800 of these were based in Poland (Tec 2003: 123). This ensured that millions of people, mostly Jewish, were exterminated.

Anfal took place on a much smaller scale, targeting a small minority of people (probably around 200,000 civilians). There were no forced labour camps or extermination camps. The mass killings happened by shooting civilians at the edge of freshly dug pits (see Chapter 1). Nevertheless, there are similarities between Anfal and the Holocaust. Civilians went through a selection process upon arrival at the camps. These selections were very similar to those of Holocaust victims where 'women with their children were one group; women were divided from men; the young were separated from the old' (Tec 2003: 121). Unlike the Holocaust, however, where the men were selected for slave labour and the majority of women,

---

32 Nalia, June 2010.

children and the elderly were immediately gassed, the men during Anfal were executed within days of their capture. The elderly were taken to a special camp in the south, where hundreds died of hunger, heat, and maltreatment, and the young families to a separate camp.

There were a number of assembling and processing centres during the various stages of the campaign.[33] Most people were processed through more than one of these centres. People from Anfals 1, 2 and 3 were processed through Suleimanya Emergency Forces, Chamchamal Liwa (brigade headquarters), Qader Karam (Elementary School and Police Station), Aliwawa jash headquarters, Laylan animal pen, Tuz Khurmatu Youth Centre, and Qoratu Fort (21st Infantry Division Base). During Anfal 4, people were processed through Harmouta Army Camp and Taqtaq Fort (Military Garrison). People from Anfals 4, 5 and 6 were processed through Spielke military post outside Khalifan, Rawanduz prison and Erbil Security Headquarters. Then, after the initial collection, almost everyone from Anfals 1-7 went through the temporary holding centre at Topazawa Popular Army Camp, near Kirkuk, where the men were last seen.

During the final Anfal, the villagers were processed through the complexes of Deraluk, Sori Jeri, Kwaneh, Sersing, Bersivi, Hizawa and Mangeshk (fort and primary school), Zakho military base and Amediya (Police Station, Army base and Teachers Union Headquarters). Then, almost everyone went through Duhok fort at Nizarka where the men were separated from the rest. A few days later, the women and children were taken to Salamya military base near Mosul, while the elderly remained in Nizarka. In the following sections I will present women's descriptions of the main camps in both the Soran and Badinan regions.

Topzawa Camp – Anfals 1-7

The whole of Kurdistan was in Topzawa. (Shekh Hassan February 2006)

There is no place like it in filth and awfulness. (Keejan, December 2005)

The camp consisted of a large one storey building, spreading over two square miles and surrounded by barbed wire. There were around 30 large halls each 25-30 meters long (MEW 1993: 209-217). The halls were brimming with people with some containing up to 250 civilians. People started coming in early April after travelling for days, and being processed through various assembling points. They arrived at Topzawa hungry, thirsty, exhausted and confused. No one knew what was going to happen to them and no one dared to ask. Anyone who asked a question or was slow in doing what s/he was told was beaten. People were off-loaded into the large, barren courtyard, surrounded by the military. The soldiers

---

33 Information about the various forts and assembling centres was gathered by Middle East Watch (1993).

were screaming orders in Arabic and they 'kicked and slapped the women and children' to speed up the offloading process.[34]

Upon arrival, the soldiers aggressively pulled the men and teenage boys out of the crowd. It was 'like wolves attacking a flock of sheep,' recalled Jwana[35] who lost four siblings and fifteen nephews and nieces to the campaign. Meena,[36] who was with Jwana at the time, reported that the men were blindfolded and handcuffed with their own jamana (a black and white, or red and white cloth, which is wrapped around the head like a turban) and pishten (long cloth which wraps around the waste like a belt). They fell on top of each other and the soldiers beat them and forced them to stand up again. Within days of their arrival they were lined up, pushed in different directions, and beaten into the closed, ambulance-like vehicles that took them to their death.

Before they were removed, the men's belongings were taken from them. Rasmia[37] talked about hills of 'men's ID cards, combs, mirrors, rosaries, knives, lighters, and clothes, all soaked in the rain.' Stripping the men of their belongings was symbolic of what was to come. From that moment on they were not individuals anymore but a bunch of frightened and powerless people who were no longer fit for existence. The confiscation of possessions and clothing, and stripping people of their previous identities also served to humiliate the rest of the detainees, terrifying them into complete submission (Tec 2003: 123).

After processing, the men and the elderly were removed from their extended families who were shouting, crying and begging the guards. The young families were herded together in the halls. The doors were locked from the outside and the windows were heavily barred. Some people were locked up for the first twenty four hours and they were reduced to urinating in the hall.[38] Some of the halls had a bucket to use as a latrine but most did not. Some women reported being allowed to go to the toilet twice a day, guarded by soldiers at all times. There was a water tank in the courtyard and the women were sporadically allowed to go and get some drinking water, depending on the guards' mood. The majority, however, were locked in without food or water. Civilians were detained in Topzawa for between a day and a week but few reported staying for longer than a month.[39]

From the moment of their arrival the women were convinced that they and their children would not survive. The halls were overcrowded and there was no space to lie down or stretch out. Everyone slept on the concrete floor with no blanket for cover. There was no systematic attempt to feed the prisoners. Twice a day the guards opened the doors and threw buns into the overcrowded halls. The food 'was attacked,' according to Shadan. She recalled that fights broke out as women

---

34   Shadan, March 2006.
35   Jwana, May 2010.
36   Meena, May 2010.
37   Rasmia, December 2005.
38   Runak, November 2005.
39   Shekh Hassan, February 2006.

tried to get some bread for their children. Those who were stronger managed to eat, while the majority remained hungry.

The prisoners were interrogated at different times throughout the day. The majority were interrogated in the large courtyard, as guards walked around and recorded people's names and details. Lana,[40] however, recalled being interrogated upon arrival in a large room where 'tables and chairs were laid out like a restaurant.' At each table, a couple of officers and an interpreter were questioning people. The civilians were asked which village they came from, what jobs they did and why they had become 'saboteurs.' When the questioning ended, Lana and her husband walked towards the door. She was carrying her eight month old son, her husband was carrying their two year old daughter, and their eldest child was holding on to her.

The couple were still hopeful about remaining together through this ordeal. They were holding hands and Lana remembered how the guards 'sniggered' at them 'because they knew what was going to happen.' At the door, they were forcibly pulled apart. Lana was so grieved about losing her husband, who was being kicked, slapped and dragged away that she forgot about her children. She was dragged to the women's hall and when she got inside 'it felt like the judgement day.' All the women were standing up and crying. At first Lana didn't realise what was going on. She thought they too were crying for their husbands. But soon she noticed that there were no children in the hall and that her own children were also missing. Then Lana started shouting again, this time crying for her children.

Various women reported their fear and grief when their children were taken from them that day. They could hear the children screaming in another hall. The lights were turned off, and the young ones were scared. Rasmia remembered how one of the guards sniggered as he told her: 'It is over, you won't see your children again.' She believed that they and their children would be 'finished off separately.'[41] Until midnight, recounted Lana, the women 'were standing up, crying and beating their breasts.' By 12.30 they brought the children back and threw them into the halls 'just like animals.'[42] In the dark no one could see eachother. Until they turned the lights on no one knew whose child they were hugging. Then the women exchanged the children and cuddled them. Later Keejan's daughter, who was eight years old at the time, told her that a Turkoman and a Kurdish guard were walking around 'pinching the children'[43] and telling them to make a lot of noise so that they will be taken back to their mothers.

Topzawa was buzzing with the people, the shouting soldiers and the various transports that brought some people and took others away. Every day, convoys of vehicles came to take people away, some to the mass graves, some to Nugra Salaman in the south of Iraq and others to Dibs. The moment of departure was as chaotic as the moment of arrival. People were brought into the courtyard, pushed

---

40 Lana, March 2006.
41 Rasmia, December 2005.
42 Lana, March 2006.
43 Keejan, December 2005.

into full vehicles and driven away. Topzawa was the last place where most of the women last saw their men.

Dibs Camp: The Women's Prison

The one storey army base was a place where the commando forces were trained (MEW 1993: 222). From mid-April women were brought to Dibs in groups of large buses and coasters. The convoy that brought Lana and her children to Dibs drove through Arab villages. She felt that they were like 'a mobile cinema' for the spectators who were 'throwing stones at the buses, spitting and swearing.' This group arrived at about 9 in the evening. A strong wind was blowing in the dark. All the lights were turned off and the women had to make their way through barbed wire. A few children were injured, falling over and being scratched by the fences. According to Lana, one woman fell onto a protruding wire and was blinded in one eye. Lana's sister, Shadan, who travelled in another convoy, also described arriving in the dark, when 'an eye could not see another.' The darkness made the process agonising. They walked a thin track through barbed wire, ascending stairs into the large halls. The next day the families were summoned to the courtyard. Shadan then found her sisters and other relatives and they stuck together throughout their time in prison.

The situation in Dibs was better than Topzawa. The families were given a blanket each and there was space to lie down and sleep on the concrete floor. However the halls were overcrowded and people lay down next to each other like sardines. There was no privacy and there was a constant struggle over space. Keejan recalled that whether you had ten children or five you had access to the same small space, roughly two square meters. Those who arrived at Dibs during the first stages, had access to proper latrines and there was no restriction on using them. Others reported shortage of toilets. In some cases a small room was used as a toilet that had no outlet. Keejan recalled a few women finding an old scythe and piercing a hole in one of the corners of the room to make a latrine. Women took turn cleaning the toilets every morning, according to Keejan. The guards gave them spades to shovel the faeces into containers which they had to empty out in a section of the courtyard.

The facilities could not cope with the number of people. The bathrooms 'were always blocked, filthy, not enough for the people,' according to Shadan. She remembered how when they used the bathrooms they had to lift up their dresses up to their knees not to get dirty. The smell from the bathrooms was overwhelming. It made people feel depressed and sick. Many people had rashes, wounds, and cold sores on their faces because of the insects and dirt. People were riddled with lice which were 'walking all over the prisoners,' stated Lana. She said that at first the women tried to maintain control by cutting their own and their children's hair, trying to kill each other's lice and hanging clothes in the sun. Soon they realised that the lice had 'more power over [them].'[44] This is similar to Bondy's (1998: 318) experience, a Holocaust survivor who talks about the war 'against dirt and insects' in the ghetto of Theresienstadt where the bugs 'resisted all efforts to keep the room clean, to air the bedding, to disinfect.' In both cases the women lost the war against bugs and filth.

Water was available but there was no soap. Most people only had the one set of clothes they were wearing. Children had diarrhoea, and young infants suffered from sore skin and rashes as their mothers hand washed their clothes with water. During menstruation the women had to wash their undergarments with water, and wait in the toilets until they were dry enough to put them back on. Sometimes the blood dried in their clothes. Vago (1998: 277) who survived Auschwitz also talks about how 'the overall filth was aggravated by the last menstrual period that all the women had.' During the Holocaust, many women stopped menstruating due to the starvation diet, violence and stress. Anfal survivors, however, talked about their suffering during menstruation. Some of them tore off a section of their dress to use as a towel which they hand washed every month. They were scared and embarrassed about hanging them on the barbed wire in the courtyard. Only children's clothes were taken out to dry in the open air. The halls smelt of sweat and blood no matter how much the women tried to stay clean.

During illness, some women were allowed to go to hospital while accompanied by soldiers while others were not. Women reported not daring to go forward, even when they themselves and their children were gravely ill. They were worried about being abused by the guards on the way (see section on sexual abuse, below). In the last month a medical group also visited the Dibs camp. Some women believed this happened as a result of their complaints,[45] while others believed that the guards were fed up with burying so many dead bodies and they took pity on the prisoners.[46]

The prison guards soon found a way to make money. They were selling soap, tea, rice and other goods at inflated prices. Those who had money managed to give their children better food. Twice a day, trucks brought food for the prisoners. They were given buns and tea for breakfast, soup for lunch and sometimes even rice and stew. Mothers tried to save a little bread to give to their children at dinner time. Some reported first giving the food to their children, and if there was anything left over then they themselves would eat.[47] Everyone talked about being hungry. The buns were so hard 'you needed a hammer to break them.'[48]

On a regular basis, convoys of vehicles came to take women and children away. Hall by hall their names were recorded, then they were transported, and they disappeared. These were the families who, like the men, were shot in the

---

44 Lana, March 2006.
45 Lana, March 2006.
46 Keejan, December 2005.
47 Runak, November 2005.
48 Shadan, March 2006.

mass graves. According to Razaw[49] and Ismat,[50] each time about 200 women and children were taken away. The guards did not allow the women to take any of their belongings with them, not even the baby's formula and bottle. They were stuffed into the closed military vehicles which had one small window at the back. Ismat reported that the children, who did not know where they would be taken, were waving goodbye and smiling at the prisoners who were left behind: 'They thought they were going to be released.'

Once, Razaw recounted, it was their turn to be taken away. The halls before them were all empty and Razaw's group would be the next to go. She is convinced that it was her mother's constant praying that saved them. For some reason their room was missed out: 'The al-mighty God blinded then when it came to us. It was as if they did not see our room.' They transported the next room instead, and her group were saved by pure chance. Each time an empty hall and the abandoned belongings of people set fear in everyone's hearts. Razaw reported that Said Yusuf, one of the Dibs guards, would come into the hall on the next morning and tell the women: 'Pray that the wolf does not take you.' They asked him why he said that and he replied, 'I know where they are taken. I pray to God that you won't go there.'

Ismat's group were transferred to Tirkrit and they were supposed to be shot at the graves. About a week after their arrival at Tikrit the infamous closed vehicles were brought. The women were called into the courtyard, their names were read out one by one and they were shoved into these vehicles. All their belongings were confiscated, including their earrings, rings, money, children's clothes: 'You will be killed, they told us, you don't need these things anymore.' The cars were crammed full of people and the doors were closed on them. They were held up, however, because one of the women in the group was not on the list. It was September and the heat was oppressive. Soon people were fainting from lack of air and the excessive heat. About two hours later, at sundown, the soldiers finally opened the doors and let the women back into the halls: 'Water, sweat and people poured out of the cars.' The next day the General Amnesty was announced and this group were saved at the last moment.

During the summer many children died because of the heat, starvation diet, lack of sanitation and illness (see section on Death of Children). Many women also fell ill. One woman died leaving five children behind.[51] Other women fell ill and those who were lucky were helped by relatives. The plight of women who gave birth in the camps and those who were ill and incapable is discussed in another section.

---

49 Razaw, May 2010.
50 ,Ismat, June 2010.
51 Lana and Shadan, March 2006.

Nugra Salman – The Elderly People's Camp

It was a hole at the end of the world. Only the unfortunate were taken there. (Gurji, November 2005)

This is Hell, they told us, you have been brought here to die. (Haji Ali, February 2006)

Nugra Salman was a two storey prison fortress in the south of Iraq. The Saudi hills could be seen in the distance. The sky was constantly hazy because of the dust. Surrounded by desert and isolated from human settlements, it was an ideal place to hold prisoners (MEW 1993: 226-231). People could try and go beyond the barbed wire but no one could escape. A woman who tried to flee came back hungry and thirsty few days later, claiming she had gone in search of wood for fire.[52] The prison held somewhere between 6,000 and 8,000 people during Anfal (MEW 1993: 226). A man recalled overhearing two prison guards saying that 1500 prisoners lost their lives before the amnesty.[53] This would constitute 20-25 per cent of the total number of prisoners. Some survivors believe that it is a miracle that they came back from Nugra Salman despite the widespread hunger, filth, illness and lice. 'Wasn't it a miracle?' Qumri [54] asked, 'Everywhere you turned there were men and women falling'.[55]

Each truck, which had space for 25 people, brought 50 elderly prisoners to the camp. The journey from Topzawa to Nugra Salman took 12 to 14 hours. Sixty-two year old Samira[56] from Mamsha village fainted in the truck because of the heat, lack of air, hunger and thirst. The soldiers stopped for food and the toilet a number of times. Samira asked for food and the other prisoners scolded her. They were scared that if they annoyed the soldiers they might be killed. Samira replied, 'They are going to kill us anyway so let them get on with it.' Surprisingly, the soldiers gave them some bread. Other convoys were not so fortunate. In one of the trucks that transported the Halabja returnees to Nugra Salman three children screamed for water every time the guards drank. Two of the children died soon after their arrival at the camp.[57]

Qumri recalled how the soldiers stole their money on the way to Nugra Salman. Just before arriving at Samawa the guards told the prisoners that if they had any money they should hand it over for safekeeping because otherwise they risk getting their money confiscated at the checkpoint. Many people naively passed on their money. When they arrived at Samawa (without being searched at the checkpoint) the guards got off and a new set of guards took over from them. According to

---

52 Shekh Hassan, February 2006.
53 Smko, March 2006.
54 Qumri, May 2010.
55 Qumri, May 2010.
56 Samira, March 2006.
57 Naji, March 2006.

Qumri those who had given their money to the soldiers 'were beating themselves in regret.'

Haji Ali arrived just before sunset. He described being surrounded by an empty desert: 'There were not trees, no springs, just dry desert land.' It was only when he arrived at the courtyard did he realise that his group were not the only ones who were brought there. He broke down in tears as he remembered the moment of realisation that 'the whole Kurdish nation was there.' He met old men and women from all the different tribes he knew, all the different regions he had visited or just heard of. Everyone's names were recorded upon arrival and they were hustled into the large halls.

Each hall had two internal guards. This consisted of more privileged prisoners who either spoke Arabic or were considered 'more civilised.' Some of the internal guards abused the other prisoners and were despised by them. One of the more popular internal guards[58] said that every 24 hours he received about 1800 buns in dirty blankets. Whatever time of the day it arrived, he had to go and receive them. The bread arrived in lorries and was then divided between the different halls. This internal guard was responsible for 400-500 people. He allocated a number of people who would then distribute the bread to their own group. Each person received 3-4 buns for the 24 hours – breakfast, lunch and dinner, and 'there was absolutely nothing else.' Early on during the month of Ramadan, however, people were given rice and stew and even meat. Soon things changed and everyone was entitled to three hard buns a day with a little container of water, 'You broke your tooth on them.'[59]

At first the prisoners were given salty water from a well in the prison camp and many people fell ill as a result. Then once a day a truck came bringing the detainees water. Each prisoner had a small container to get her water ration for the day. Sometimes when they queued to get water, putting their containers in a line, they were randomly beaten by the soldiers. Sinab from Masoyee village in Qara Dagh put her container before the water hose when one of the guards picked up a thick hose and hit her on the hip. She lost consciousness for a few minutes and had no water to drink that day. Gurji recalled how people rushed for the 'sweet water tank' every morning. Once when somebody knocked over her container in the queue and she went to straighten it up she was hit by a cable on the shoulder. She was 'desperate for days, unaware whether it was day or night.' Shekh Hassan was fortunate enough not to have to queue for water. He along with five other male prisoners were given a room which had a water tap, 'God had made it easy for us.' Access to water was a valuable thing and he managed to stay clean and well when others were 'filthy and sick.'

As in the women's camp the toilets were overflowing with filth and many people suffered from diarrhoea. The prisoners themselves cleaned the toilets. Those who had money paid others to do it in their turn, others just had to wrap

---

58   Shekh Hassan, February 2006.
59   Gurji, November 2005.

their faces and get on with it. Survivors believed that the diarrhoea was brought on by the salty well water because some people who were desperate in the long hot days drank it anyway and fell ill. Shekh Hassan recalled how everyone smelt awful, 'the smell of shit and sweat' and how the lice were 'climbing people like ants.' Qumri remembered how some days she would wake up and her black dress was white as if 'yogurt was poured on it.' This was because of sweating in the night and the desert dust that stuck to her damp clothes.

Shekh Hassan recalled how the ill, those who 'shit themselves, who were disabled by illness and parasites' would be thrown out to the courtyard.[60] The large desert flies were circling them and 'their faces were covered with mucus.' They would reach out from the hole in the wall and beg people for food and water. Some people felt sorry for them and gave them a sip of water or a slice of bread but most of the time they were left to die in the heat. Qumri believed that it was the heat that killed people, they were put under a brass cover by the wall. Some of the people who were taken out to the courtyard 'weren't that ill, the scorching sun killed them.'

Various people reported being beaten by Hajjaj, an Arab lieutenant who was renowned for his cruelty. He beat people by hose and cable, he stood on their backs with his boots, tied them up to poles in the heat, and he whipped them. Sometimes he beat people until they 'became loose' or until he himself 'got tired.'[61] Qumri recalled how Hajjaj punished three men who had made a hole in the corner of one of the toilets to provide an outlet. He first made them lie down and roll around in the toilet filth and then he tied them up to electric poles in the blazing heat. The men were screaming and crying from pain. After a few hours, when they were finally taken down, 'the skin of their backs was left on the poles.'[62] Two of them died soon after that.

Sebri[63] was beaten by Hajjaj three times during her detention in Nugra Salman. Once he complained that they had not kept the hall clean and he ordered everyone to get into the oppressive heat as a punishment (something which he regularly imposed on prisoners for arbitrary reasons). Sebri quietly sneaked back to bring her children's ID cards in fear that they may not be allowed back into the hall. She was bent over to pick up the ID cards when he started beating her with a cable. He beat her all the way to the courtyard, hitting her on the head, spine and shoulders. Another time he beat her because her children were lying down in a shady part of the courtyard. Finally, on a very memorable occasion the prisoners were called into the courtyard to be given a slice of watermelon each. There were far fewer slices than there were prisoners. Sebri was told to sit cross legged on the floor. He then started beating her saying that she was not sitting properly. She kept turning to different directions as he was beating her, trying to sit properly but whatever

---

60 Shekh Hassan February 2006.
61 Qumri, June 2010.
62 Qumri, May 2010.
63 Sebri, May 2010.

she did he kept hitting her. Sebri has chronic headaches which she traces back to Hajjaj's beatings: 'There is a continuous whistle in my head. Sometimes when I am busy I forget about it but it is always there.'

Every day a number of people died in the various halls. Shekh Hassan who took part in burying the dead pointed out that at first they used to wash the dead, wrap them in cotton (bought from the prison guards) and bury them according to Islamic custom but soon there were too many dead people and they just gave up. Those who were alive were exhausted, hungry and weak. They could not keep up observing the rituals of burial and mourning. The graves were shallow because after digging a few inches into the sand you would reach the rocky bottom.

A widely circulating story is that of the black dog in Nugra Salman that dug up dead bodies and ate them. Various men and women reported seeing the dog with dead people's clothes in his mouth, the clothes they were buried in,[64] and some reported seeing the dog chewing children's arms and legs.[65] This caused intense fear. Nasik[66] talked about a woman who spoke Arabic, translating for the other women what one of the guards was saying. According to Nasik the guard said: 'I pray to God that you don't die here, otherwise the black dog will eat your body.'

This story, however, was disputed by Shekh Hassan who took part in burying the dead. He reported checking on the graves and finding that they were untouched. Whether or not the story is true it terrified the detainees in Nugra Salman and continues to haunt them and their children long after Anfal ended. Children who were born after Anfal reported having nightmares about the black dog that ate the dead children.

Forty year old Naji who arrived in August was stunned by what she saw. The family had returned from a refugee camp in Iran under the illusion of an amnesty (see above). When they arrived one early morning the prisoners came to the windows to watch their arrival. They were 'like animals locked behind bars (from the second floor). Their hands were hanging out of the windows, their hands were black.' Naji wondered to herself whether these blackened, skinny and broken figures were human: 'They were looking at us from the windows but some didn't even dare to look at us properly.' Most people were so dehumanised by this stage they had no will of their own. By the time they were released their skin 'was black like the clothes [they] wore.'[67] This is consistent with what survivors of the Holocaust experienced upon their arrival at camps where others had already been detained. Rita, a Hungarian Jewish survivor recalls arriving at Auschwitz when she met 'people who look[ed] crazy' because they were pale, skinny, had no hair, and they were wearing strange clothes (Tec 2003: 120).

---

64   Meena, May 2010.
65   Qumri, May 2010.
66   Nasik, May 2010.
67   Gurji, November 2005.

Duhok Fort in Nizarka and the Women's Prison in Salamya

It was the place of the dirtiest Baathists, the place of Istikhbarat (intelligence), the place of torture. (Adiba, February 2006)

The fort is a large, two storey concrete building with a courtyard in the middle. Thousands of people from the Badinan region were detained there. As was the custom in Topzawa the men were immediately separated from the rest of the villagers, blindfolded and handcuffed. Sometimes the women had to wait in the courtyard until space was found to move them to. They stood by and watched as the men were beaten. According to some people a number of men were killed by being hacked in the head with concrete blocks on the day of their arrival.[68] Ameena[69] arrived at the fort with her three children after the amnesty was announced. The courtyard was empty at the time but bloody concrete blocks were scattered everywhere. She asked one of the Kurdish guards why the blocks were bloody. 'This is a butchery house,' he told her. She asked whether there was a butcher in the fort and he sniggered, 'they slaughter humans here.'

The halls were full of waste. The women, according to Adiba, had to brush away the faeces with the edge of their hands to make space for sitting down. They were given neither food nor water. The toilets, for those who were allowed to use them, were blocked and overflowing. Most women stayed in the fort between two to five days. The elderly, however, were kept here until the amnesty. The women with their children were transported in small buses and IFA trucks to Salamya in Mosul. They remained there for two weeks until the General Amnesty was announced.

Adiba from Darkari Ajam had a baby and a toddler. When she queued up to get on the bus to Salamya the soldiers took her baby's 'milk and bottle out of [her] hands.' She believes that they 'intended to let [them] die.' The baby cried 'all the way to Salamya' and she eventually 'died of hunger.' The babies who were breastfeeding were slightly better off. They fell ill but they survived. Most of the others who drank formula milk died because they were 'deprived of food' and 'their intestines dried up.' Adiba herself had no milk to give to the baby. She had not eaten for days. Unknown to her she was also at the early stages of another pregnancy. She felt unwell and lethargic. Her baby had diarrhoea because of the heat, hunger and filth. Some days she hand-washed her daughter's clothes 15 times with water.

As was the case in Dibs, Salamya was an improvement on Nizarka. In the large one storey building women had access to water although no soap was available. They could come out to the courtyard and wash their clothes. Some reported being given food while others spoke about hard buns that were difficult to chew. When the amnesty was announced the women were asked to get up and dance because

---

68 Adiba, February 2006
69 Ameena, February 2006.

'the merciful leader had forgiven [them].' Fairooz[70] refused to join the soldiers until she was struck on the shoulder with a cable and forced to stand up and dance. Women cried as they danced with their prison guards. Some of them didn't even believe that this time the amnesty was for real.

Women's Social Experiences in the Camps

Sexual Abuse

Addressing sexual abuse was the most challenging aspect of this research. Asking questions about possible abuse raised suspicion. Most women looked alarmed and sometimes a little scared. I usually tried to ask more sensitive questions after knowing the women a little better. This is because in Kurdish society, rape brings disgrace on the victim. The rape victim may be stigmatised, blamed for being raped, forced to relocate to a different region, deprived of custody of her children and even killed by her own relatives. Women are held responsible for the crimes committed against them. This may be one reason why they remain silent about these issues. Most of the women I interviewed did not speak about sexual abuse during the Anfal campaign or after the amnesty. Some of them fiercely denied having even heard of anyone being raped. They knew that if they admitted that such things took place, they themselves might be considered a possible victim. Others would go as far as saying it is possible, or that they had heard of such things but had never witnessed it. Sexual abuse never happened to them or to their relatives and acquaintances. In two cases, however, women spoke about sexual abuse spontaneously.

I have been conscious of the potential danger in exposing issues such as rape and abuse. I finished my fieldwork in Iraqi Kurdistan and returned to Europe, where I started transcribing and analysing my interviews. At times I have felt frustrated when important issues had been hinted at and I had not followed them up. Once I asked a woman whether the guards sexually abused the women. She replied, 'No, unless a woman was horny and asking for it.' I now keep thinking: Was she hinting at prostitution in the camps? Were these women who were 'asking for it' trying to get extra bread for their children? Who could be 'asking for it' when hunger, dirt, illness and death were widespread? But I did not ask her these questions and we moved on to a different issue. Before I started my fieldwork, I was prepared for encountering difficulties when it comes to women talking about sensitive issues, but I was unprepared for my own cautious attitude. At times, if a woman used 'metaphoric language'[71] about abuse and circled around the subject, I too found it difficult to ask her a direct question about it. I felt that I should not press them on issues that they found embarrassing or difficult.

---

70 Fairooz, February 2006.
71 Das (1997: 84).

I visited one of the Garmian villages on various occasions. It was only after several trips and a long process of trust-building that some of the women told me about a young woman in Nugra Salman who was loathed by other prisoners because she allegedly had a relationship with Hajjaj.[72] As in other cases 'the valuable commodity' in this exchange was 'not gold, diamonds, or money but food' (Bondy 1998: 320). The young woman was called by Hajjaj, whom everyone feared, and she returned few hours later with extra food for her family. One woman remembered that sometimes Hajjaj called the young woman intimately as if he was calling his wife to sew a button for him. After their release, this woman changed her name and moved to another area to avoid reprisals. She is hated by members of the community who know her.

Some factors interfered with conducting thorough interviews. First of all, in most cases I was unable to speak to the survivors alone. Usually, when I visited a house many of the villagers would come there once they heard about the purpose of my visit. The village houses are small and it was difficult to find the space to talk alone. When I told women that I wanted to interview them alone many said, light-heartedly, 'There is no need, we keep no secrets from each other.' Sometimes I interviewed women in the presence of men and women from the village, but mostly the women sat together in a separate space to talk to me. In some cases I felt that the women wanted to tell their story in front of other women relatives. Roshna who spontaneously talked about her separation from her sisters-in-law and her night in solitary detention, cried and told me that she had never told her relatives about what happened (I will return to this issue later). I felt that she was trying to tell her two sisters-in-law as much as to tell me. The interview seemed like an outlet for her to talk about a subject that hurt too much but was not allowed public exposure.

Speaking of rape by the Soviet occupation soldiers in the immediate post war Germany Grossmann (1999: 177) points out that by the time the men returned the women had already had their abortions and treated their venereal disease. The women were then 'left with memories that had not been worked through, that had no easy access to public space even as they were ... constantly invoked or alluded to.' Similarly, although rape during Anfal is mentioned in the public discourse, the survivors have been unable to speak about their experiences.

Throughout the 1990s there were reports about women being found in Kuwait, Saudi Arabia and Arab parts of Iraq who were allegedly sold to Arab tribes during Anfal. To find out about the possibility of women being sold I spoke with Adalet Omar who worked on documenting Anfal for six years.[73] She pointed out that news about such women had been confirmed from several sources. First of all, a member of the Popular Army who worked in the Topzawa camp was a witness to the selection of some girls. The man testified about beautiful girls being handpicked

---

72   The name of this woman is undisclosed as this is socially sensitive information.

73   Adalet Omar, Adviser to Ministry of Human Rights, Refugees and the Anfals. Erbil, April 2005.

and taken away by a group of men who had an official document that sanctioned their behaviour. Secondly, in 1991, a Kurdish soldier found his sister in Kuwait. She had disappeared during Anfal, was sold on the border of Kuwait and ended up marrying an Arab man and having two children with him. Her brother brought her back to Kurdistan with her two children but the woman refuses to be interviewed. I have often wondered whether this woman's return and her silence were her own choice. Butalia (1997: 101) talks about the forced recovery of abducted Indian and Pakistani women during the partition where 'women's wishes were of no consequence.' It was the State that decided what was best for the women and returned them to their previous communities even if they had converted to a new religion and had formed new families.

On occasion silence is imposed on women. When I was visiting the Garmian region, which fell prey to the third Anfal and from which the largest number of women and children disappeared, I was told about a woman who had been initially in the same selection group as the 18 women mentioned in the secret Iraqi document which was later published by the Kurds.[74] This woman is now remarried and her husband does not allow her to give interviews about what happened to her and what she knows of the others.

More recently, after the fall of Saddam's government in 2003, members of the Arab tribes of Shatra and Shanafian from the south of Iraq made contact with the previous Ministry of Human Rights (Erbil) to report about Kurdish women who were given to them as 'presents'.[75] Adalet Omar who represented the Ministry visited these tribes but she was not allowed to meet the women. The Ministry offered to take the women back to Kurdistan but they were informed that these women are now mothers, and have settled down, therefore there is no need for such a thing. The tribes merely wanted to officially inform the Kurdistan Regional Government to avoid future tribal reprisals from the women's families. Significantly, the women themselves were not allowed to talk to the visitors and to say what they wanted. After talking about several examples that have come to the attention of the ministry Adalet Omar then concluded: 'Whether that particular document [about the 18 girls] was authentic or not there is a strong and terrifying voice which tells us that Kurdish women were either given away as presents or sold to Arab men during Anfal.'

From my own research, I found that men were more willing to talk about women being abused than the women themselves. A middle aged man who was detained in Nugra Salman camp on the border of Saudi Arabia was the only person who reported witnessing a rape. He was a clean man, cautious not to let the lice and flies weaken him. He protected himself from dirt and neglect. The toilets near him were 'all blocked and dirty' so he used the other toilets which were further away from his hall, but cleaner. When he turned the corner he found that a soldier

---

74   Top secret Iraqi document reveals Kurdish girls sent to harems and nightclubs in Egypt, 7/2/2003. KurdishMedia.com.

75   Adalet Omar, Erbil, March 2006.

was raping a woman and, noticing him, 'he quickly took himself off her' and walked away. The woman was 'tall, young, and beautiful.' It is possible that the woman had taken her child to the toilet when the soldier caught her alone: 'She was bleeding, that is how hard he had been raping her.' The woman who was weeping abundantly said to him: 'What life is this?' His only answer to her was that 'God knows everything.' He did not stop to console her or ask questions, he tried 'not to make it worse for her.' He told me that this woman is still alive but he would protect her identity until he dies. This was only the second time he had spoken about it. For a long time after his release he believed it best not to repeat such stories. He thought talking about them would not make anything better.

The threat of rape was more readily discussed by women. Bana, who was a strong and outspoken woman, was brought up in a small town and had studied up to 1st year secondary school (equivalent to the 7th grade in the British system). Unlike most of the village women she spoke a bit of Arabic. She was arrested with her husband and two children. They were taken to the temporary holding centre in Tuz Khurmatu where the men were separated from the women. She pointed out that the moment they were separated from the men 'the soldiers came to have a look at [them].' Unlike the other women she could understand them commenting about the beautiful women, joking between themselves and laughing: 'I swear if I could have died, I would have died right there... The other women didn't understand; they were naïve.' The soldiers 'laughed between themselves, they were saying rude things.' Bana then took initiative and advised the women to 'bring [their] scarves down to [their] foreheads and look down.' She spent six days in Tuz Khurmatu and despite the fact that people 'were rotting' in dirt, sweat and tears, they 'didn't try to clean up.' She advised the women that it was best to smell.

Another woman talked about arriving at Chamchamal holding centre. She had two young children and was pregnant. She watched from the coaster bus how hills of Kurdish clothes, combs, mirrors, rosaries and ID cards had been piled up. The helicopters were circling above them; crowds of confused people were being pushed in different directions. But she remembered a particular image which she found embarrassing and possibly a hint at the awful things to come. From the abandoned clothes she saw a woman's undergarment (which looks like Aladin's trousers) spread out on the front of a military truck, the legs opened, as if it was a woman who was spread to be raped. Nadia talked about a similar 'humiliating and shameful' experience in Topzawa. The soldiers put a woman's undergarment on a stick and raised it in the air. 'Hay allam Akrad,' they shouted: This is the Kurdish flag.

Only one woman, Ahoo, talked about the real dangers of being raped and the relationship between the women and their prison guards. She was detained in Dibs with her sisters and their children. Unlike her older sisters, who denied fearing rape, she talked about this issue. The prison guards were initially members of the Popular Army whom she believed were good people. Later they were changed to the *mughawiri* force (commandos) who were 'immoral and without conscience.'

They tried to have sex with the women, first by seducing them and telling them that if they consented, they would have 'a better life.' Then, they started threatening the women that 'if you don't have sex with us we will not give you food.'

Ahoo was writing letters to her comrades outside the prison and bribing a guard to pass on her letters to a contact outside prison. She was found out and because of these activities she was transferred to Kirkuk Security Office for further interrogation. During investigation, they kept saying that she is doing peshmarga work. She explained that she was only trying to defend herself and her fellow prisoners. However, she had been a secret member of the PUK (Patriotic Union of Kurdistan) in the past and was imprisoned twice before Anfal. She had a file which marked her as a 'saboteur.' Eventually, she confessed that she had been a peshmarga and would carry on in that path. She believed that things could not get worse – her husband had disappeared, her son had starved to death, and she herself was a prisoner. One of the interrogators warned her, however, that 'there are worse things,' and by this he did not mean death.

Ahoo admits to being tortured. When I asked her about the torture she was quite vague about it: 'kicked and slapped and things like that.' She was detained in Kirkuk for three nights. She was beaten by a cable: 'torture and things.' I did not ask her whether she was raped because all the way through talking about this she was looking down or looking away, not wanting to be stopped. Later in the interview, this woman told me how after her release from prison, her parents-in-law took her children away from her and would not allow her to see them. Although she would not specify the reason for this, it was obvious that people assumed that she had been raped during her interrogation in Kirkuk. This made her 'a bad woman' who could not be a good mother. For longer than ten years she met her children 'in other people's homes.' Ahoo was brave in addressing the issue of rape but she was not prepared to talk about her own victimisation.

A fellow researcher, who interviewed Ahoo on the same day, told me that she had confessed to him about being raped in Kirkuk. This puzzled me for a while. I kept wondering why she would tell him and not me. There may be different reasons for this. First of all, I interviewed most of the women on camera, unless they refused to do so, whereas my colleague was taking notes. It would be easier to deny such subjects when there is no audio or visual recording. Secondly, I had already interviewed her before he came and maybe repeating the story and the interference of what Langer (1991: 6) calls 'deep memory' had something to do with this. Another factor may be the gender dimension. I have naively thought that it is easier for women to talk to other women about intimate and sensitive subjects, yet this may be an incident which proves otherwise. It may be more difficult for some women to speak about being raped to another woman. On the other hand, Macek[76] believes that when communicating with each other, women do not necessarily spell things out because certain things are assumed to be understood.

---

76 Ivana Macek, The Uppsala Programme for Holocaust and Genocide Studies, personal communications.

It is possible that a woman believes another woman is more likely to understand what really happened without having to explain everything, whereas a man may need to be told more directly.

The problem we face when researching about sensitive issues is lack of sufficient information. The question I ask myself is: What do I do with information disclosed by one person and not confirmed by others, not even her own sisters who lived in the same camp with her? Does this mean the information should be disregarded and that the Dibs guards presented no threat to the women? How do I know that this is reliable information? How can I find the truth in what is based on one person's memory? Does the witness remember correctly? Is she telling the truth? Is she exaggerating? Or should I be asking myself different kinds of questions: Why do the others not talk about sexual abuse? Why do they deny it? Is there a group agreement about silence to protect the name and reputation of everyone? Have women been advised to be quiet? Najmadeen Faqe Abdullah[77] who was one of the first people to research about Anfal told me that just before being released from the camps, the elders had advised the young that whatever happened in the camps should be abandoned there. This advice was particularly directed at women; don't tell anyone about your own particular suffering or that of another woman, this is for the best.

It is possible that a story, told 18 years after it has happened, may have changed, some aspects ignored and others highlighted. But it is important to include these stories. Leydesdorff et al. (1996: 5) point out that feminists and oral historians have come to the recognition that the subjective dimension of 'the hidden voices' of women and other oppressed groups is fundamental to understanding them. The subjectivity of oral sources has come to be seen as a strength rather than a weakness. In other words, narrative promotes a new, accurate understanding of objective conditions that derives from personal experiences yet transcends the boundaries of the individual.

Lack of support for a particular story does not necessarily imply that it is false. This is particularly true about sensitive matters such as rape, theft, prostitution, and generally what is considered 'morally wrong.' The one witness who speaks about these issues may be telling us about something important that the community wants covered up. For the purpose of my research, I have no choice but to rely on this kind of information because most women are silent. In the context of Anfal and the other sources about missing Kurdish women, it is very likely that the guards did present a threat to the women prisoners.

We also need to be aware that even after 18 years these stories do not lose their potential danger. Women may be still vulnerable to further stigmatisation and abuse. Women's silences may not always be an oppressive strategy forced on them. In some cases, this may be a rational choice made by women who want to protect themselves from the social consequences of such disclosure. It is also possible that some women still find it too painful to talk about these issues. There

---

77   The Committee to defend Anfal people's rights, 1992-1994.

is evidence that some people find talking about their experiences therapeutic (Aron 1992: 175) but this does not mean that everyone finds it so. During my PhD research with Kurdish women refugees and my Anfal research, I met many women who did not want to talk. Talking about the past can bring it alive once more, and this may be something that many women are trying to avoid. Levine (1995: 84) points out that 'The survivor who allows memories long buried to be brought forth again risks renewed pain, a return to anguished memories which they endure alone after the researcher has turned off the tape recorder and the interview completed.' Hence, only the voices of those who were willing to talk have been included in this research.

Roshna, as mentioned above, was a willing speaker who talked about being 'separated from the others' with great pain. When she started telling me about this event, her sister-in-law, whom she was sitting next to, mumbled something to her. She replied, looking at me: 'She [meaning me] will find it useful.' She was not silenced by this attempt. In Dibs camp her son, who was a toddler at the time, was starving to death. She was watching as some women went out and came back with food (another hint at prostitution). She assumed they were buying food from the guards. She put her *aba* round her and went to one of the guards. She showed him her money and asked him for bread. Her son was 'about to die.' The guard gestured that she followed him. She followed thinking that he would give her food from the storage. He took her to a small room and locked her in. She was terrified.

Roshna did not say what exactly happened to her in that cell but she kept crying and saying that it was the most awful thing. She even wished that she and her children had starved to death but this 'terrible thing' had not happened to her. Only later when I switched the camera off I asked her whether people assumed that she had been raped. 'People are awful,' she replied. Following this she kept insisting that she is a 'good' and 'pure' woman, that no one has touched her and her conscience is clear but 'people talk.' In other words, she equated being raped with becoming 'impure' and becoming 'bad.' Part of me wondered whether she brought this up in front of her relatives to tell them how difficult it had been for her and to explain to them how it had happened without her choice. 'I was naïve,' she kept saying, 'I didn't know anything.' She cried throughout this story, leaving us to imagine what happened to her. Part of me also wondered whether she is worried about her second son who is now a married man. I wondered whether she was trying to defend herself in front of her relatives who may whisper bad things into his ear.

Neither of these two women talked about being raped, but the language of pain and grief implied it.[78] Roshna seemed to have ached about her experience in silence for many years. Ahoo has suffered gravely from the consequences of her interrogation and torture in Kirkuk, losing her children to her parents-in-law. However the dangers of rape are more profound for unmarried girls. These

---

78   Other researchers have been able to get direct disclosures (see Omar 2007) but I can only speak about my own sample here.

women are unable to marry because they are not virgins any more. After the fall of Saddam's regime in 2003, three women made contact with Adalet Omar to seek help. A few years after their release from the prison camps they started getting marriage proposals which they could not accept. Their families had been pressurising them to get married and they did not dare to say why they could not: their secret would not be safe anymore. Adalet visited the families accompanied by experienced social workers. After a long process of negotiation they managed to disclose this information to the families and now the young women have found husbands amongst their own relatives. But how many more are still suffering in silence? How many have been forced to marry and then killed because they were not virgins? Those women's stories will never be told.

Researching on sensitive issues such as rape is never an easy task but it is particularly difficult in a patriarchal society that, on the one hand, uses the issue in its political discourse to justify its turn towards nationalism and, on the other hand, victimises women who are known to have been raped. Women who were raped live with the burden of silence and the fear of being found out. It is no wonder that during the Anfal trials the Kurdish community managed to convince no woman to talk about being raped, although complete immunity and confidentiality was offered to those who would speak. As a result rape was not included in one of the charges against the perpetrators.

Women instinctively negotiate their way through sexual violation and its echoes in their lives. They must silence their anger about the injustices they have endured and try to lead normal lives. If they continuously fall ill, feel depressed and suffer from repeated headaches they can always blame the genocide campaign for taking away their fathers, husbands, brothers, and sons. However, in reality, they cannot blame Anfal for their rape. They cannot admit to having been raped because to do so would damage their reputation and jeopardise their lives.

Birth and Loss of Children in the Camps

One of the issues that interested me in this research is the story of women who were pregnant, those who gave birth during the campaign and the problems they may have faced as new mothers while nursing a baby on a near starvation diet. Surprisingly, some of the women felt that these issues were not worth mentioning. No one spontaneously spoke about periods, infections, labour, birth and nursing. Even when I asked questions about these things many gave short and general answers. This goes to support Ringelheim's view (1997: 20) that gendered memories are considered irrelevant and gradually forgotten. I also found that the theme of birth was regularly discussed alongside death by the women. For many women witnessing birth was followed by witnessing death and this led to the psychological association of the two together. Death of children was a prominent subject in most of the interviews, even for those who had not lost a child themselves.

During the Holocaust giving birth or having young children was in itself a death sentence. Young mothers were immediately sent to the gas chambers with their children. They were considered useless as a labour force. Babies and children were also considered the continuation of a group which was singled out for annihilation. Himmler when talking about giving the order to kill Jewish women and children explained that although this was not an easy option he did it anyway because he could not allow 'avengers in the shape of children to grow up.' (Goldenberg 1998: 327). Women helped each other give birth in secrecy during the Holocaust. In some cases, they immediately killed the baby and told the mother it had been stillborn to protect her from being sent to the gas chambers (Goldenberg 1998: 329). In some ghettoes women were forced to have an abortion when they found out that they were pregnant. Those who refused were transported to the gas chambers (Bondy 1998: 314). Anfal babies were not automatically killed in this way but many of them died as a result of wilful neglect.

When they were being transported to Topzawa in IFA trucks, a woman went into labour in Keejan's truck. It was raining and the people were all muddy, hungry and tired. As the truck drove along the bumpy country road the woman gave birth to a boy. She was weeping. Her husband had just been taken away from her and then she had a baby in an overcrowded truck surrounded by strangers. The woman didn't want the baby and she threw it out of the truck. 'What can you do with a baby in the dirt and rain?' Keejan asked, 'When you have nothing to put on him and you cannot feed him, when you are dead?' Keejan still wonders sometimes, if that woman is still alive, how does she feel about throwing the baby away? Does she cry for the baby who was never given a chance? Does she feel guilty? Does she wish she had kept the baby? Keejan still struggles with similar issues and questions. She stays awake many nights and suffers from repeated headaches. Death would have been easier, she kept repeating throughout the interview, but 'when the soul does not depart, what can you do?'

Two women went into labour in Keejan's hall in Dibs. One of them gave birth and she and her child survived, while the other woman died with the baby inside her. It was an agonising death which many people witnessed, including the children. The guards watched from the window, but did not allow the woman to go to the doctor. The dead woman was then thrown aside, Keejan's daughter recalled. She lay there until the guards took her away the next day.

Runak witnessed a number of births in the different halls and many of the babies did not make it. 'I saw it with my own eyes,' Runak stressed, 'the mother had no strength to get up. She had no milk to give to the baby and no one to help her. She lay in bed till her baby died.' Runak herself had a baby boy in Dibs. She begged some women from the Erbil district for 'a piece of cloth' to wrap around her son. After the baby was born she was 'depressed' and she stayed in bed for a long time 'as if in a coma.' She had no appetite and no wish to live anymore. Fortunately her son survived.

Women who gave birth were taken to a separate room, if available, or else the children were let out of the room while the women formed a circle around the labouring woman. Meena went into labour one evening, at the early stages of her detention in Dibs prison. She recalled the sound of the rain on that cold night. There were over two hundred women in the hall; many of them complete strangers: 'I didn't know what to do from embarrassment.' Her relatives moved her to a toilet where she could give birth in privacy. The lights were switched off by the guards (as they were every night) and Meena's mother and aunt had to burn scraps of paper and plastic to help the delivery. When the baby was finally born a few hours later, they used a blunt pocket knife to cut the umbilical cord. Meena was then shivering from cold and exhaustion. Her mother, Hajar, lay down on top of her and covered her like a blanket because they had nothing else to cover her with. Hajar also ripped a piece of her undergarment and put it around the baby.

Shadan, who witnessed a number of births in the camp, recounted that there was no doctor, no medication, no help and 'It is just God who helps you survive at such times.' The women, according to Shadan, used broken glass, rusty knives (smuggled in or bought from the guards), and even stones to cut the umbilical cord.

Saadia started going into labour in the truck which brought her to the Duhok fort. She was a beautiful sixteen year old woman. It was twilight by the time they had sorted people into the different groups and put them in separate halls. That night there was a blackout. Her relatives begged the guards to allow her to go to hospital as this was her first baby and the fort was 'overcrowded and full of shit.' Surprisingly, the guards took her to a hospital in Duhok but they did not allow anyone to accompany her. She was taken in a military truck, surrounded by soldiers. Two of them held her arms and she was crying. She cried because she was scared of the soldiers, of giving birth, of losing the baby and for her husband whom she had just seen beaten and humiliated.

The soldiers insisted on being present during the birth arguing that she may try to escape. The doctor gave them his own ID card and promised them that if she runs away they can take him to prison instead. It was a difficult birth and the doctor said that she may need an operation. She begged him to try for a natural birth because she was going back to prison. She gave birth to a boy and the soldiers did not allow her to give him a Kurdish name. She named him Muhamad but has always called him Kovan. At the registration desk in the hospital they asked for her and her husband's ID, she told them that they had none. The soldiers started swearing at her and her baby saying that 'the bastard is a saboteur.' Saadia was then taken to Salamya in Mosul where she stayed for another two weeks before the amnesty was announced.

Many babies died because of infections, lack of sanitation, and their mother's inability to feed them. The children died from hunger and epidemics. Diarrhoea, measles, chicken pox and malnutrition were common causes. Shekh Hassan, who was detained in Nugra Salman, helped bury the dead in that camp. Each day 'five or six people died' in the halls near him and most of the dead were children. At first they buried the dead immediately but as time went by, they 'just got fed up' with the number of corpses especially because they had no energy to dig the floor properly: 'Many of the corpses that stayed around for a few days were babies and young children. Their mothers hugged their little bodies for a few days before we had a chance to bury them.'

Lana who was detained in the women's prison in Dibs was one of the first women to lose her baby son. He was still breastfeeding and she had no milk. Hunger and grief at losing her husband made her 'dry.' 'Many many children died,' recalled Lana, 'A woman had five daughters and brought back one. My sister also lost her son. He was seven or eight, he fell ill. We created a graveyard, the graveyard is still there. It was just for Kurdish children.' She also pointed out how after a while the death of children became an everyday occurrence for the prisoners: 'A child would die over there and we were like mad, we were laughing and singing. Everyone had this psychological state; no one cared about the other.' This made her feel guilty later, she had flashbacks about the dead children and continuously questioned herself, How could we laugh?

Habiba also lost a baby son in Dibs. He was a beautiful healthy baby that the prison guards in Dibs loved and cuddled. The driver of the bus that transported the family from Topzawa to Dibs had a large tin of formula milk which he had bought for his own baby. Habiba begged the driver to give her the formula for her son who had not eaten for a few days. The driver agreed and this helped the baby survive for a while. Three weeks after arriving at Dibs, however, Habiba's son caught the measles from other babies. She did not dare to request a visit to the hospital. Her mother-in-law advised her that as she was young and beautiful she should not go with the soldiers. Her son died 'in an awful way.' By the end he was reduced to skin and bones and 'he cried until he had no strength to cry anymore.'

Keejan lost three children to Anfal as well as her husband. Keejan's eldest daughter was eleven years old. She was 'very beautiful.' On the night when the children were separated from their mothers the guards singled her out from amongst her siblings and took her away. Keejan is convinced that her daughter must be one of the many beautiful Kurdish girls who disappeared and were allegedly sold during Anfal (see section on Sexual Abuse). 'Just as they might walk into this room,' she told me, 'they might not like me and my daughter (who was in the room with us) but they might like you and if they do they will take you.' That is how she imagined the selection had happened. They chose the pretty girls and took them. 'Kurds kill for honour,' she said, 'but over there we had no power, we were nothing.'

Keejan's three year old son fell ill in Topzawa. He fainted in her lap while they were locked in the overcrowded hall. She knocked on the door and begged the guards to let her go to the courtyard. She wanted to get to the water tank and give some water to her thirsty son. At first the guards refused her request but she kept knocking and begging until one of them let her out. She walked up to the tank and sprinkled her son's face with water when a familiar voice from the men's hall called out to her. It was her brother-in-law who was also her cousin. He stretched his arm out of the window giving her an empty cheese tin. He begged her to give him a sip of water. He had not drunk any water for three days. Unknown to her, a soldier was watching Keejan from behind and listening to their conversation. She took the tin from him and was crouching before the water tank. Her son was on her lap when the soldier attacked her from behind. He started kicking and slapping her and her son jumped from her hand and landed face down onto the tap. He bled from his nose and mouth and died a few days later. She believes that he died as a result of an injury from the iron pipe.

Keejan's third child, a six year old girl, died in Dibs because she suffered from epilepsy (see also Chapter 6, section: Depression). Until the next day Keejan cuddled her daughter's dead body. The guards came to measure her for a grave. Keejan kept her child on her lap until the next morning when two soldiers came and gathered all the dead bodies to take them for burial. She was allowed to accompany her daughter. The guards had made a mistake on measurements and the grave was too small for her daughter. They bent her legs to fit her inside. Keejan asked the soldiers to make the grave a little bigger because she was worried that 'the wild dogs might dig her out.' The soldiers started swearing in Arabic and quickly covered the girl's bent figure. They blamed her for the wrong measurements. There was a separate cemetery for the dead prisoners in Dibs which Keejan used to visit regularly for years after her release until the security situation deteriorated in the aftermath of the second Gulf War. Two of Keejan's children are buried there.

The Factors that Helped Women Survive

Despite imposed restrictions and powerlessness women were not passive in the face of detention. Some factors helped survival, even though people had different experiences. Some feminist scholars of the Holocaust have argued that helping each other during the Holocaust was an important factor that kept women alive (Ringleheim 1998, Goldenberg 1998). Jewish women, it is argued, coped better than men in the concentration and labour camps because they bonded and supported each other. Since women are traditionally brought up to be nurturers they were better equipped 'to create and recreate 'families' and so provide networks for maintenance that may be related to survival rates' (Ringleheim 1998: 379).

According to Bondy, who was an 18 year old detainee in Theresienstadt, women tried to convert their new bunks in the ghettoes into a home (Bondy 1998: 311) and they were good at helping each other (Bondy 1998: 319). Goldenberg (1998: 327) talks about the women's memories of the Holocaust which 'emphasise women's strong concern for one another as well as their dependency on one another to withstand the barbarism of the camps; their adaptation of homemaking skills into coping skills; and the effects of their heightened physical vulnerability and fear of rape.' The author talks about the strategies women used to enhance their own and their fellow women prisoners' chances of survival. The memories Goldenberg looks at emphasise the significance of 'connectedness, nurturance, and caregiving' (Goldenberg 1998: 336). The formation of groups, it is generally argued, encouraged women and sometimes actively forced them to survive when they were too ill or too hopeless themselves (Ringelheim 1998: 383).

In a reappraisal of her earlier work, however, Ringleheim (1998: 384) points to some serious problems with her previous research which amount to her 'unconscious use of cultural feminism as a frame through which to view Jewish women survivors.' She stresses that instead of criticising the social norms and expectations that construct gender differences cultural feminism essentialises these differences. Men and women, according to essentialists, are radically different and women's culture is superior to men's culture. Hence, Ringelheim (1998: 386) argues 'femininity', which was considered oppressive in the past, is made 'sacred.' The author then goes on to state that:

> ... the discovery of difference is often pernicious because it helps us to forget the context of these supposed strengths- oppression- and to ignore the possibility that they may be only apparent. To suggest that among those Jews who lived through the Holocaust, women rather than men survived better is to move toward acceptance of valorisation of oppression, even if one uses a culture and not a biological argument. Oppression does not make people better; oppression makes people oppressed.

Similarly, Bos (2003: 27) stresses that the emphasis on women's strengths and their bonding which enhanced their chances of survival led to a 'generalised (and sometimes essentialised) analysis of gender difference which glorified women in general instead of cataloguing the historical experiences of individual women. Certain events and experiences described in memoirs and interviews that were perhaps unique or even exceptional were thus subsumed into a model which was meant to apply to all (Jewish) women and which suggested a (sometimes not so) subtle idealization of women's strength.'

Unlike the Holocaust, where men and women were held in concentration and labour camps, men were taken from the temporary holding centres straight to the execution site during Anfal. They did not survive long after their capture (most of them were killed within days) so their coping strategies cannot be compared with that of women's. It is important, however, to look at the factors that helped women survive Anfal. The women I interviewed mentioned money, being helped by others and luck as factors that helped them to survive.

Access to 'money' and 'help' were reported as two main factors by Anfal surviving women. Those who had money were more likely to survive because they could buy food from the guards at inflated prices. Seher[79] survived Dibs because she had managed to smuggle money into the camp. She was detained with three sisters and their children. Together they were 20 people. They helped each other survive. Two of her nephews died however, as a result of 'fever and sickness.' Samira[80] and Qumri, two older women who were detained in Nugra Salman also

---

79   Seher, March 2006.
80   Samira, March 2006.

stressed that they survived because they had money. A cup of sugar or rice was worth one dinar, which was equivalent to three dollars.

Runak is certain that she only survived because her relatives and acquaintances from Qawali looked after her. After giving birth she was too weak and too depressed to get up. For a while she was not even conscious. Her relatives 'poured water into [her] mouth' and covered her when she was cold. She believed that only relatives helped each other, no one else really could afford to care about you because 'they were all desperate themselves.' Keejan said that as time went on people became more desperate and 'because of their own misery people did not care about others.' A woman lost three children in one night and 'no one cared,' Keejan recalled. Everyone was preoccupied with the business of their own survival. Gurji, who was detained in Nugra Slaman also denied that people helped each other: 'If someone had money they would buy food and eat it. If they did not they would starve. No one helped each other.'

People's experiences were different but most of those who reported being helped were assisted by relatives and friends. Few people were aided by strangers. Sometimes, however, relatives abandoned their own kin. A woman from the elderly people's camp abandoned her disabled sister in another hall. Her older brother was also ill and incapable so she made a decision to help him survive. Once she visited her sister and took a kettle of juice to her. The woman was in a really bad way, being fed and taken to the toilet by others, she looked a mess. She had scratched her head so much 'her hair was sticking up in all directions and her head was bleeding.' The lice and flies were 'all over her.' She begged her able bodied sister to take her to her hall but this woman replied that she could not, stating that there was no space. It is possible that she had made a rational choice about who to help and she believed that her brother's chances of survival were better than her sister's. She admitted to not being able to look after everyone at once. She had to look after herself too. When the amnesty was announced both her siblings were alive, but as crowds of people queued to be released she grabbed her brother's hand and took him forward. She left her sister behind for others to look after.

Keejan was helped by her relatives from outside the camp. Her brother in-law, who lived in a district and was not Anfalled, had become friendly with one of the Arab guards in Dibs. The man was bribed to smuggle money and goods into the camp for some of the women. His name was Hadi and he was blind in one eye. One day he came into the hall to fix the light bulb and he told Keejan that her son looked like her brother-in-law. She denied knowing anyone of that name. Throughout her stay in prison Keejan denied having any male relatives other than those who were already captured, because she was worried that they might also be rounded up and killed. The man then told her that he knew she was related to her brother-in-law and he has sent her some stuff: 'It was twilight, it was getting dark. He too was scared of his superiors,' Keejan recalled. Hadi gave her a package which contained a letter and some clothes and shoes for the children. In the letter her brother-in-law told her that he had sent her 700 dinars (equivalent to $2100 at the time), hoping that she had received it. Keejan had never received the money.

Other prisoners who had claimed to know Keejan, had taken the money from the guard and never passed it on to her. 'We are a really self-eating people,' she said. Sometimes, the ordinary moral expectations broke down. People were reduced to resorting to prostitution, theft and 'selfish and unethical behaviour.'

Some women reported forming groups to help other prisoners. In the Dibs camp where some of the guards were harassing the prisoners, three women formed a group to advise and help the others. They all spoke Arabic, having grown up in small towns and having received some education. The group made some rules for the women to stick together, never to take children to the toilets alone, and to try to look unattractive. In the night these women slept with their shoes on. They never felt secure enough, and felt that they should be ready in case something happened. Once they managed to rescue a young woman, who had been lured into an empty hall at the end of the corridor by a guard who promised to help her escape. Some people managed to avoid sexual harassment by observing certain rules laid down by the prisoners. Most of the time, however, whatever people did and however much they tried they could not protect themselves from death and especially from losing their children.

Conclusion

After days of travelling with minimal food and water civilians arrived at the camps exhausted and unaware of what was awaiting them. Women experienced progressive powerlessness at the hands of a large bureaucratic system that uprooted them from their homes, arrested them near the main roads, sorted them by age and sex, and processed them through the long maze which was Anfal. All the while they were unaware of what the next step would be, where they would be taken, and whether they would be reunited with their divided families. Living through the Anfal camps was characterised by fear, loss, hunger, exposure to dirt and disease, and encounter with violence through witnessing death.

Eighteen to 22 years after Anfal, women still find it difficult to talk about their vulnerabilities in the prison camps. In this patriarchal society the issue of sexual abuse and victimisation could be used against the victims, blaming them for what they could not have prevented. This was not true in the case of physical abuse and beatings. Women talked about this kind of pain and fears without reservation. When it came to questions of prostitution, rape, and sexual harassment, however, there was silence. This silence was at times imposed on victims by a society that does not want to know and is not going to help. Sometimes, however, this silence was by choice, protecting not only the individual victim but also the name and reputation of the group, Anfal surviving women, as a whole.

Many women lost children in the camps and some of them gave birth. Giving birth was an agonising ordeal. Women reported feeling ashamed as they gave birth in the presence of others in the hall, many of them strangers. The umbilical cord was cut using rusty knives and razors, broken glass, and stones. Some babies died because of infections, lack of sanitation and clothes, and their mothers' inability to feed and look after them due to exhaustion and hunger. The spread of epidemics such as measles and chicken pox, were also determining factors. Women who had complications during birth did not make it. In the end the death of children was so widespread that people could not feel grief anymore, a condition which made them feel guilty later.

Yet camp life did not amount to passivity. The women quickly learned to cope with their new situation. Relatives stuck together and shared the limited resources they had, and informal networks were created to warn other women of potential threats. Individual women, however, reported different experiences of cooperation and support during camp life. Generally, those who had relatives and friends were better supported than others who were more isolated. There were times, however, when complete strangers helped each other 'for God's sake.' Women who had been sick, who gave birth in the camps, and who were in a depressed mood, reported that they only survived because of help provided by friends and relatives. Some of them are still surprised by their own survival, particularly because many others, sometimes people who were stronger than them, did not make it. In some cases, they believed it was down to pure chance, or God's wish, that they were not deported for execution at the mass graves and that they did not die of hunger and illness.

Chapter 3
Forcibly Displaced Civilians

The Anfal offensives led to the mass movement of people internally and across Iraq's borders with Iran and Turkey. Many people from the first Anfal in the Jafati valley and some from Anfals 5th, 6th and 7th in the Valley of The Lesser Zab managed to seek refuge in Iran because of geographical proximity. Similarly, during the final Anfal in the Badinan region some people escaped to Turkey. Alongside those who migrated beyond the borders thousands of civilians secretly sneaked back into the towns and cities and they became internally displaced. Some of them were rescued by friends and family when the campaign was in full swing, others escaped with help from the jash forces, or through their own initiative by taking chances and being lucky.

Iran and Turkey are signatories to the UN convention on refugees.[1] There is, however, a geographical limitation on Turkey's protection and only persons originating from Europe will be granted refugee status.[2] Non-Europeans can only have temporary asylum in Turkey and it is the responsibility of the UNHCR to find more durable solutions for them. Asylum seekers may live in limbo for long periods while waiting for refugee status determination (RSD) interviews. Throughout this period they have no access to work or social assistance and their movements are restricted.[3] Civilians are more likely to be granted asylum in Iran. This, however, is usually a long and bureaucratic process. Refugees face restrictions on their freedom of movement, have limited access to the labour market, and have to pay for medical treatment. People can move out of the refugee camps but this usually means that they would be deprived of state support and they would have to look after themselves.

According to the UN Convention a refugee is

> an individual who owing to a well founded fear of being persecuted for reasons of race, religion, nationality, membership of a particular social group or political opinion, is outside the country of his nationality and is unable or unwilling to avail himself of the protection of that country. (1951 UN Convention Relating to the Status of Refugees)

Internally displaced people (IDP), on the other hand, are

---

1 Islamic Republic of Iran on 28 Jul 1976 (Convention & Protocol) and Turkey 30 Mar 1962 (Convention) 31 Jul 1968 (Protocol).

2 http://www.unhcr.org/cgi-bin/texis/vtx/page?page=49e48e0fa7f.

3 Ibid.

persons or groups of persons who have been forced or obliged to flee or to leave their homes or places of habitual residence, in particular as a result of or in order to avoid the effects of armed conflict, situations of generalised violence, violations of human rights or natural or human-made disasters, and who have not crossed an internationally recognised State border. (Guiding Principles on Internal Displacement, 1998, Introduction, para. 2)

The main difference between refugees and internally displaced people is that IDPs, unlike refugees, do not cross national borders even if they have fled their homes because of conflict, violence, and human rights abuses. This in turn means that, as citizens, they remain under the protection of their own government, even if they have been forced to flee by the same government. The majority of civilians who were internally displaced by Anfal were hiding from the Iraqi government until the September General Amnesty. For this group, displacement was accompanied by constant fear of being found out, arrested, and possibly killed.

Uncertainty about the future and concerns about their current socio-economic status are central to the experience of forcibly displaced individuals (Grant-Pearce and Deane 1999). The violent disruption of routine goes hand in hand with exposure to new, harsher, physical and social conditions. Loss of control over the environment and the inability to avoid problems or prevent them undermines self esteem. Parvanta (1992) stresses how change to socioeconomic status makes the individual feel vulnerable and depressed. A man may lose his job and therefore his role as the breadwinner and he may experience loss of self esteem which in turn will affect his relationship with his wife and children. The author also places great importance on the loss of support that women receive in their daily lives back in their home countries. She argues that such support is neither attainable in the country of exile nor replaceable.

This chapter addresses the experiences and impressions of refugees and internally displaced individuals during the campaign. The intention is to draw out a range of issues that are important when working with uprooted individuals. These include traumatic experiences such as surviving danger, witnessing death and violent destruction, the hardships during escape, and the problems during resettlement and/or hiding. These in turn affect the wellbeing of exposed individuals. Issues related to safety and meeting basic needs must be tackled alongside awareness of the displaced people's vulnerability and traumatic past.

Refugees

Forced displacement is the result of escalating violence. The process is identified by three phases. These are pre-flight, flight, and resettlement (Ager 1993, Veer, 1998: 9). The first phase is when the person experiences increased oppression in her country before she is personally targeted. This involves the social and political changes that affect the whole community. These events become part of the community's history, narrated to explain and justify migration. The second phase is the period in which the individual personally becomes a victim of terror and the deprivations experienced while escaping in order to seek asylum in a safe country. The third phase involves life in exile. This is when the person is experiencing the after effects of her past traumatic experiences, the uncertainty of future, the problems of adaptation, and the possible experiences of racism and discrimination.

During the first phase as violence escalates in the home country, individuals feel more vulnerable and less in control. Finally, the threat to one's life becomes imminent and staying is not a viable option (the second phase). From the early 1980s, the Iraqi government had lost control of the villages which were targeted by Anfal in 1988. The inhabitants called this region 'the liberated zone' and the government called it 'the prohibited zone.' In this zone people lived with the fear of bombardments and attacks but they were able to carry on with a reasonably normal life. They planted their own land and were financially independent from the government and the peshmarga. They used the peshmarga hospitals when they needed medical care and their children were taught in schools that were independent from the government. Even the bombardments had become part of the everyday routine and people learnt to live with them. This is why many people were astonished at how destructive, brutal, and bloody the campaign turned out to be. The scale of the destruction and massacre was incomprehensible for most people because the attacks were unlike any other they had witnessed in the past.

Nian,[4] who narrowly escaped freezing to death as she escaped via Kanitoo (see below) reported that 'no one believed that it would be as it was.' When the planes came to gas Ware and smoke descended on the village, Layla[5] was still not concerned: 'We could not believe it because there were no peshmarga in our village, there was no (anti government) activity.' Keejan[6] reported that even though there were rumours about a large government attack they did not leave Bangol village because, 'no one believed that they would do this to us.' Nask,[7] from Koshk village in Garmian, stated that 'no one believed the government could do it.' Fairooz,[8] from Derkari Ajam in Badinan, stated that even when they were arrested, people did not believe they would be killed because 'how could 182,000 people get killed?'

The failure to comprehend the change in Iraqi government's policy towards the Kurds left many defenceless. The senseless and over-arching violence caught everyone by surprise and they did too little too late to protect themselves. Living through the Anfal campaign was characterised by witnessing destruction, death, and sometimes a direct threat to one's life. Every woman in this research has

---

4   Nian, March 2006.
5   Layla, April 2006.
6   Keejan, December 2005.
7   Nask, December 2005.
8   Fairooz, February 2006.

lost members of her family, extended family, or village. Everyone's homes were levelled, their farms were burnt down, and their animals and belongings were looted. Many of the women were in direct danger and survived by pure chance. In the following sections I will describe some of the women's journeys to safety (the second phase of the refugee experience) and their situation in the refugee camps and towns as they resettled in Iran and Turkey (the third phase).

The Journey to Safety

I will present three kinds of escape situations during Anfal. The first Anfal of Jafati valley started in February, at the heart of a bitter winter. I will describe the refugees' escape journeys in the snow as they walked the tough mountains to the border of Iran and were then transported by Iranian vehicles to safety. This journey took 4 to 7 days during which various people died by freezing to death or falling down mountains and bridges. The second example is from the region targeted by Anfals 5, 6, and 7. The inhabitants of Warte escaped gassing and conventional bombardments and walked to the border in late spring. They were hungry and thirsty and it took them about one month to get to Iran because of taking detours to avoid the army. The third flight situation is from the region of Badinan, targeted by the final Anfal. These civilians fled large scale gassing and many of them were injured. They also faced another difficulty as the Iraqi army was attempting to block their escape route to Turkey and then pursued them all the way to the border.

Escaping Anfal 1 in Winter

In early March, when the peshmarga were losing the battle in Jafati valley, people started preparing to hide in the mountains. As the attacks were proceeding, one rainy noon in the village of Kanitoo Nian, a writer who with her husband had fled to the peshmarga controlled region in the early 1980s, was preparing food with other women to take to the caves. A young man put his head through the door and told them to run because 'they (the army) are coming.' This caused huge chaos and 'everyone started fleeing at a moment's notice,' recalled Nian. It was already midday and the rain was pouring, creating streams in the valleys: 'It was a strange thick rain, one drop after another, you could not see a metre ahead of you.'

People were trying to cross the streams, some were shouting for help and others were swearing. By the time people reached the mountain there was a snowstorm because, according to Nian: 'in the cold country if it rains in the valley then it is snowing in the mountains.' People started climbing the mountain, going against the wind and snow: 'The snow was hitting you, you kept slipping and falling, you were cold, shivering … The mountainside was very steep, we had to walk one behind the other. If your foot slipped you would fall down and no one would help you get back on your feet. The rough hills and icy valleys were below us. You had to tread very carefully.' A woman who was walking behind Nian was carrying a two month old baby. Her foot slipped, she fell, and screamed once. A man told the woman that, if she could, she had better get up because no one was going to help her back on her feet: 'We all saw her roll down into the valley, no one did a thing to help her.'

Soon people started freezing as they walked in the snowstorm, unable to see each other. They were bombarded by 'clumps of snow that kept hitting your ear, your face, all over.' Nian recounted that people started throwing away whatever they carried in order to be able to walk faster: 'I saw peshmarga throwing away their blankets, coats, clothes. Some even threw away the pictures they had in their pockets, they thought that they would be lighter and they could walk faster.' A wounded peshmarga, who could not walk anymore, slipped into a sleeping bag, zipped himself up and said goodbye. When she walked past him he was already being covered by the snow. That day more than eighty people froze to death in the space of a few hours and the rest of the people gave up and started retreating.

Initially a young man was carrying Nian's one year old daughter. After about an hour he came and gave her daughter back. 'She is freezing to death,' he told her, 'I don't want her to freeze in my arms.' The toddler was not conscious anymore. Children who were being carried were more at risk of freezing because they were not moving and their bodies got colder more quickly. Nian carried her freezing daughter along until she fell into a ditch and the baby leapt from her arms. She remained in a ditch 'full of snow and slush' until an acquaintance came and grabbed her out. She was disorientated and she did not even look for her daughter. Much later when Nian retreated to the bottom of the mountain a young man came carrying the little girl. They thought she was dead but as her father kissed her goodbye and was preparing to bury her he realised that a 'tiny breath was left in her.' They put her near the fire and massaged her face and body until she could cry once again. For the time her daughter was missing and later when they thought she was dead Nian didn't cry or feel anything: 'A number of things happened in a few hours,' said Nian, 'things you could never imagine. At such times you are detached from your normal feelings, these things happen and you become abnormal so whatever happens during this time seems normal.' According to Herman (1997: 34) encounter with extreme danger can change ordinary perceptions and make people disregard hunger, pain, and exhaustion. This mobilises the threatened person's energies for strenuous action, flight in this case. However the emotional numbness Nian felt at the time, made her feel guilty later (see Chapter 6).

People retreated back to one of the unoccupied villages and had to take a different route on the next day. There were 'ant lines of people' walking in the snow. People were happy, Nian recalled, 'amazed to see each other alive. You are alive, they said, You have not frozen.' On the second day they arrived at a village which was high up a hill. She saw 'masses of people, like a raging sea.' On the other side of the village there was a flat ground where there was shelter. The Iraqi planes bombarded the fleeing refugees and Nian is still amazed how no one died as a result. That night they stayed in an abandoned village close to the border of Iran. The women took refuge in one of the houses but soon it was filled up with fleeing villagers from Sargallu, Haladin, Chalawa and other villages in the region.

Nian and her friend did not sleep. There was no place to lie down and the two women chatted and laughed about their current situation. This was their way of coping with what was happening to them. Her friend joked that thanks to Saddam Hussein, who keeps destroying their homes, they have to make a new one every so often. Unknown to them an older woman nearby was listening to their conversation. At dawn the older woman's son came to fetch her. The two young women cheerfully asked her whether she was leaving because this meant they would have more space to stretch out. She glared at them with anger and pity: 'Yes I am leaving. May you get mouldy here. For Quraan's sake, it is as if you are going to be wed.' Nian laughed recalling their own naivety and the irony of the situation. The old woman's last words were more out of pity than anger: 'Poor children! You are going into exile, you are going to Iran. You will be exiled and displaced but all evening you have been talking and laughing.'

The next day, they carried on with their journey, getting stuck in the mud and slush, falling into ditches, falling behind and catching up with each other. Sometimes, the paths by the mountainsides were too narrow. People walked one after the other and if one person stopped, the whole line would come to a standstill. The villagers walked for two to four days until they reached the border. In Galala they were picked up by Iranian cars which took them to Iran. The bridge that refugees needed to cross in Shanakhse was blown up by the Iraqi planes that continuously bombarded the region. A temporary bridge was put together by the Iranian soldiers. It was made of wooden boxes and rope. The ropes were tied to the iron construction of the old bridge on either side of the river. There was a long queue to cross the bridge. It took Nian's family three hours of waiting before they were able to go over and continue their journey by car. Crossing that bridge was the hardest part of her journey, Nian recalled. Each step was a struggle as the unstable wooden boxes shook under her feet and she had to hold tightly to the ropes. 'I shouted to all the saints and sheikhs and prophets until I crossed that bridge,' Nian laughed.

After crossing the bridge the families were picked up by Iranian cars once again and taken to a refugee camp on the border where they were given food and blankets. The family then moved on to Mariwan where they stayed with an acquaintance for ten days. It was here, when the stressful situation was over, that they all fell ill: 'You know that sense of arriving,' Nian stated, 'we started falling ill... We had a terrible cough, the family [we stayed with] could not sleep because of our coughing violently through the night. The cold was starting to show its effects.' Finally, they were told to move on to Seqiz where they rented a house and where many of the peshmarga families were gathering.

Other inhabitants of the Jafati valley left via Jhilwan Mountain. Although some people froze to death, died as a result of exposure to the gas, hunger, and exhaustion, there was no large scale loss of lives as in Kanitoo. Dalia[9] from Chalawa village walked to Iran carrying two of her children on her back. Her parents-in-law also

---

9   Dalia, April 2006.

carried one child each. Some of the villagers in the region left their elderly and disabled relatives behind. They could not carry them along and they believed that the elderly would be immune from murder. Later, they realised how wrong they were when they found the bodies of their dead relatives shot by the mountainside. Some of the elderly who had remained in the village were taken to Nugra Salman camp and some never returned from there.

The civilians walked the tough Jhilwan mountain amidst 'two meters of snow,' recalled Dalia. Amongst the snow and slush some people lost their shoes and they were too numb to notice it. Others who stopped off to warm up by a fire got their clothes burnt. As they tried to cross the Shanakse bridge Dalia's 6 year old niece (her brother-in-law's daughter) fell into the river stream. She was 'taken away' by the rushing water and they could not even stop to watch her die. Younger children were carried over the bridge by Iranian soldiers who were more experienced in using such bridges. Iranian cars picked people up at various points and Dalia sometimes managed to get a lift while other times she kept walking until she got lucky again. People did not dare to stop in fear of being gassed and killed. Later they were gassed in the Iranian refugee camp near Bana when Iraqi planes dropped their bombs there. Four people died as a result of this attack and many others died in the following years (see below and also see chapter 4).

Escaping Anfals 5, 6 and 7 in the Late Spring

On 23 May, the planes dropped up to eight gas bombs in Warte, two of which fell into the river. Two were dropped in the middle of the village and up to four in the back side near the mountain.[10] Few people were around at the time which is why only six people were injured. The majority of the inhabitants were already living in the valleys and caves nearby, tending to their animals. Many of the men were in the 'backing force' and they were fighting alongside the peshmarga. On the eve of the gas attack women, children and the elderly abandoned the valleys and fled to Karoukh mountain. Only an old man and his sister, Qadir's[11] father and aunt, remained in the village. They believed that the government would not harm them. The old man's grandson returned to the village to persuade him to flee: 'He said that the government is not so brutal to burn us and burn our house on our head,' recalled Qadir. 'He thought the government had conscience but there was none.' The two old people were then arrested. Their house was burnt down as they watched. Their cattle and sheep were looted and they were trucked away to Nugra Salman. They were never seen again.

Dozens of families hid in the caves and under large boulders. Karoukh mountain was bombed by planes and cannon from the morning until dusk. The villagers did not dare to move because of the continuous shelling. A number of women with

---

10   This was confirmed by various villagers from Warte such as Qadir, Dara and Gulle.

11   Qadir, March 2006.

their children had taken refuge in a cave while their husbands fought alongside the peshmarga. In the afternoon a rumour spread that the government was planning to seize the mountain. Some villagers gathered their children and decided to leave their hideouts and flee. Gulle's[12] eight year old daughter was standing at the mouth of the cave when a shell dropped close by and its pressure lifted her into the air. She fell into the valley below: 'When she fell she shouted, O mum! That was all.' Gulle recounted, 'At that moment I tried to catch her but I didn't manage. Just like a bird flying off suddenly, that is how she went.' Gulle wanted to jump after her daughter but people stopped her. They tried to convince her that her daughter may be alive but Gulle did not believe them. The mountain was very high. By the time they went down and found her, she was dead. Her body 'was hid in a boulder hole, covered by leaves and dirt.'

The families then walked to Iran. The journey took about one month because the government had blocked all the roads and people had to hide and negotiate their way through the mountains. The jash helped people escape by guiding them through the difficult regions and providing transport in some instances. All the villages they passed through were deserted and there was nothing to eat: 'Wherever we found a place to rest, we took a rest.' Gulle recalled, 'It was mountain after mountain all the way to Iran.' The families arrived at Pishtashan, a beautiful and prosperous village which was deserted. A few peshmarga were still hanging around. The children were screaming from hunger: 'We had been so rich and then we had absolutely nothing to eat,' Gulle remembered. They were exhausted, hungry and in ragged clothes. The peshmarga gave them some food which helped them along to make the rest of the journey. During the long march through the tough mountains a nine-month-pregnant woman fell and she was in severe pain for the rest of the journey. When she finally gave birth in Iran 'her child was born disabled.' Gulle grieved for her daughter and she kept crying. Most people could not sympathise with her because it was not yet clear whether they themselves might survive the attacks. It was only after arrival in Iran that 'we started talking about who had died and who was left behind.' Only then were they able to mourn those who had died.

Escaping Anfal 8 in the Summer

During the final Anfal, when more than a dozen villages were gassed, people headed to the Turkish border. The village of Birgini consisted of 15 extended families with the population reaching 100 people.[13] On 26 August, just before sunrise, 8 planes dropped their gas bombs on the village. 'They came when the children were still sleeping,' recalled Mihammad.[14] Unlike the past when the bombs made a loud noise, these bombs were rather quiet. During impact a white and yellow

---

12   Gulle, March 2006.
13   Mihammad, February 2006.
14   Mihammad, February 2006.

smoke was released which quickly spread through the village. It smelt like apples, garlic and orange. At first it smelt good but it immediately made people short of breath. Mihammad had learnt that the best way to fight chemical weapons was to wrap wet towels around your head and to build a fire. He shouted from the top of his voice and warned the villagers that it was chemicals and they should seek protection. Four people from different households who did not have instant access to wet towels died immediately. One of the victims was Nalia, a particularly heavy woman who weighed about 100 kilograms. She was still breathing when the villagers left. They abandoned her near a spring after pouring water all over her. Her husband and children marched with the rest of the villagers, no one was capable of carrying her.

All the sheep and cattle died immediately, about 5,000 in total, according to Mihammad. A yellow liquid was oozing from their open mouths. Some people managed to put wet towels on their mules and horses, knowing how vital they would be for escaping. The families then started their exodus towards Turkey. In the chaos and panic Mihammad's family left behind their money and gold, too eager to leave the region behind. As they marched the villagers started feeling 'dizzy and ill.' Their skin was blistered, their feet were swollen, and some people went blind: 'Our bodies were shaking. A person's head before our eyes would become twenty heads. One tree became ten. We put our foot on the floor and our foot slipped.' When they reached the village of Zinava, the Birgini survivors were given milk and yoghurt which made them throw up. Some of them believed that it was this that saved their lives because 'it cleaned up the poison' inside them.

The inhabitants of Birgini walked for twenty four hours until they arrived at the main road which is parallel to the border of Turkey. The Iraqi troops were already gathering, trying to block the refugees' path to Turkey. The Birginis, along with what seemed like 5,000 other villagers to Mihammad, found an outlet between Sivie and Hizawa. Thousands of people were gathering on the other side of the main road, close to the border of Turkey. Another four people from the village of Birgini also lost their lives by the time they reached the border. One of them was Mihammad's seven months old daughter who didn't even have the strength to cry, she just made a weak whining noise. Her mother tried to give her some milk but she could not swallow and her eyes were closed.

It is estimated that somewhere between 65,000 and 80,000 people fled to Turkey during the final Anfal (Physicians for Human Rights 1989). When the first civilians reached the border they found the Turkish border closed. The guards stood by, not allowing anyone entry. Mihammad recalled how at the time 'the friendship between Turkey and Iraq was very strong. There was great brotherhood between Saddam and the Turks. So they did not give us entry for three days.' Ahmad[15] from the village of Barchi who also arrived at the border of Turkey around the same time confirmed that the Turks would not allow them in. Meanwhile the Iraqi army was closing in on the refugees and the planes were circling overhead. The two men

---

15 Ahmad, February 2006.

offered different reasons why eventually the refugees were allowed into Turkey. 'We knew how merciless the [Iraqi] government was, we knew if they get to us they would annihilate us,' recalled Ahmad, 'We knew what our fate would be with the Iraqi government. We had to try our luck with the Turkish government.' After a brief discussion between the village men, according to Ahmad, it was decided that they storm the border putting the women, children and elderly in front of the men. Eventually people were allowed to get through and hence saved.

According to Mihammad, when the Iraqi army arrived at Sindi gorge and there was a distance of about 1.5 kilometres between them and the fleeing refugees, some of the armed village men started shooting at the planes and the army. The Iraqi army shot back but this was too close to the border and the Turkish border guards were as much at risk as the refugees themselves. This, Mihammad argued, was the reason why the Turkish guards grudgingly opened the border to the refugees. Whatever the reason, after up to four days of waiting the civilians were allowed in. They were told that they must not make false allegations about Iraq using chemical weapons, according to Mihammad, who was moved from one hospital to the next to hide him from journalists keen to meet gas survivors. 'No gas was used against [them],' the refugees were told. They surrendered their weapons to Turkey and were allowed to cross into safety agreeing to the government's terms and conditions.

In summary, people who made it to Iran and Turkey in 1988 were escaping grave dangers. They were being gassed, bombed, shot at, pursued by the army, and some of them were exposed to extreme temperatures, hunger, and thirst during flight. They were all in direct contact with death and some of them experienced a threat to their own lives or the lives of their children. There is evidence that exposure to such traumas cause long-term mental health problems (Silove et.al. 2000: 4, Herbst 1993, Roberts et al. 2009). Herman (1997: 34) points out that certain dimensions of trauma are likely to cause more harm such as being taken by surprise, feeling trapped, over-exposure to danger, physical violation or injury (for those who were wounded or injured by gas) and witnessing grotesque death (such as death by gassing). These experiences, the author suggests, cause terror and helplessness. Postero (1992) points out that individuals who leave their own countries behind and undertake long, exhausting journeys to the country of refuge are unprepared for the 'long process' of seeking asylum. Below, I will present some women's experiences in the refugee camps.

The Refugee Camps in Iran and Turkey

Since the Kurdish parties had been an ally of Iran during the Iraq- Iran war, Iran was more receptive to refugees compared to Turkey. The villagers were picked up in cars from the border and driven into Iran where they were taken to refugee camps. Dalia, from Chalawa, along with her children, extended family and many others from the Jafati valley were taken to a camp near Bana. It was called Haware Khol. They were located in tents which were too cold during the spring. After the difficult and dangerous journey many civilians, especially women and children, fell ill because of the cold, hunger and exhaustion. This was particularly difficult because of lack of appropriate health care. Roberts et al. (2009) point out that in camp situations 'being ill without access to medical care, narrowly escaping death and being seriously injured' are associated with worsening physical health. In the weeks that followed some children died as a result of illness, exposure, and exhaustion in the absence of medical care, according to Dalia.

As summer approached, the tents became unbearably hot, Dalia recalled. The women cooked using oil heaters which made the tents and the whole camp scorching. The camp was overcrowded and there was hardly any space between each tent and the next one. This left people with no space and no privacy. At first, there was little food which was not distributed properly but things gradually became more organised. Dalia described their lives in the refugee camp where the dominant issues were hunger and fear. Living in such overcrowded camps people feared for their lives, because of the poor sanitary conditions, minimal health care, little or inadequate food, and bad security. Children could not go to school, there was no work, and people were always scared of more gas attacks.

Baron et al. (2003: 255) point out that during an emergency primary needs such as food, water, accommodation and medication take priority over mental health issues. This is because lack of a basic standard of living can be as damaging, if not more, as past experiences of violence. Women are particularly vulnerable in refugee camp situations and may be subjected to violence including sexual violence (WHO 1995, cited in Baron et al. 2003). Yet no one was willing to talk about such issues (for silence about sexual abuse see Chapter 2). The lone women, those who had lost their husbands during Anfal or before, were vulnerable to food deprivation. Rebwar[16] from Chalawa insisted that people helped these women 'for God's sake.' Yet when I asked him whether anyone fared badly he replied: 'Oppression always happens at such times, whether you like it or not but there was help too. It wasn't as if people didn't care about each other.' This may be partly true but it may also be an attempt at presenting a more humane picture of life in the refugee camp and to show that despite the difficult situations the general human expectations were met.

Refugees from the final Anfal had to rely on their own resources to survive the first few days after arriving in Turkey. Some were helped by fellow Kurds from Turkey. Others were eating the livestock they had brought with them. Those who had money could buy food. Extended families were crammed into small tents without blankets or pillows. Many people suffered from diarrhoea and some died from illness as a result of the intensive heat, change of diet, exhaustion, lack of clean water, and change of environment. Asia[17] from Sger village nearly lost her aunt and sister because of diarrhoea. Asia's family arrived at Giza where they stayed for three days. Soon rumours spread that the Turks would separate the men

---

16 Rebwar, April 2006.
17 Asia, February 2006.

from the women (something which the Iraqi government was doing in the prison camps). Many people, recalled Asia, quickly started giving their daughters away for marriage under terrible circumstances. Girls who were old enough to be married were given to their cousins in fear of what might happen to them otherwise. This is an indicator of the families' fear of dishonour and shame in the face of possible sexual abuse in the camps. Even though no cases of sexual abuse were reported, the fear of such abuse was more readily discussed.

The refugees were continuously moved around. Asia's family left Giza, their initial point of entry, where they were 'surrounded by soldiers' and walked for one day till they reached Qadisha where the nomads lived. After living there for a few days buses were brought and the refugees were transported to Gavar. The family slept under a small tent, on the bare floor and without bedding. They leaned on each other in the night to keep warm and to support each other. After a couple of weeks the Turkish government fixed up toilets in the camp which helped prevent the spread of more disease. Asia's family stayed in Gavar for forty days. They were into September by then and the weather got much colder. After the amnesty the family decided to move to Iran (see section on Migration from Turkey to Iran, below).

The inhabitants of Barwari Bala (Higher Barwar) marched towards the Turkish border after the gas attack on the villages in the region.[18] They first arrived at Biadre, where they spent one night, and then they were transferred to Gavar where the government provided tents for the refugees. This group of families decided to stay put after the amnesty. In November 1988 they were moved to a camp near Moushe and the families lived in shared accommodation. The camp was surrounded by barbed wire and Turkish guards. Here they remained until the 1991 uprising. At first the families were not allowed to work but gradually people were sporadically allowed to get out of the camp. According to Hissen 8,500 people from Badinan lived in this camp. Kurds from Turkey helped the refugees by giving them work. They hired them as farm labourers. Everyone had to be back in the camp by 3 in the afternoon.

Many children died in the Moushe camp, according to Hissen, because of the harsh winter, hunger and lack of sanitary water and medical care. None of the children were able to go to school and they had no means of entertainment. Many refugee mothers still remember their children's deprivations with great pain. At the time, when they were concerned about more urgent issues such as safety and making ends meet, they had no time to think about these things. Now that they are back in their home country, when the situation is stable and there is reasonable prosperity mothers watch the new generation of children who have access to schools, toys, artistic activities and games. They feel sorry for their own children who grew up without much of a childhood.

---

18   Hissen, February 2006.

Continuing Threat

Crossing Iraq's borders did not amount to escaping the government's wrath. Refugees were followed into their country of refuge and attempts were made to annihilate them. Here I will refer to two incidents in Haware Khol camp in Iran and Mardin camp in Turkey. Haware Khol was too close to the border of Iraq and the refugees were still fearful of the government. Soon they were proved right. In July 1988 the Iraqi government gassed Haware Khol camp. The refugees saw two Iraqi planes approaching and many of them headed to the shelters nearby. They thought that it would be a conventional bombardment. Rebwar who did not find space in the shelter was badly injured. Most tents had oil containers outside which they used for cooking and for oil lamps. With pressure from the attack two oil containers near Rebwar were lifted into the air. One of them hit his shoulder and exploded, setting him on fire and making him blind in one eye. He was also affected by the gas which rapidly spread through the camp. He survived despite being injured 'by two chemicals (gassing and petrol).'

Four people died instantly as a result of this gas attack and many were wounded. In the weeks that followed others also lost their lives in hospitals in Bana, Tabriz and Tehran. Dalia and her children inhaled the gas. She had not witnessed a gas attack before and she was surprised that the explosion had no shrapnel and was so quiet. Soon the smell which was not unpleasant at first 'blinded' them. Many people lost their sight when the day got hotter and their glands became swollen. They could not swallow or open their eyes, according to Dalia. At first the Iranian guards did not believe that it was a gas attack but the refugees knew too well that it was. The injured were first taken to Bana hospital which was a small hospital without appropriate facilities. They were then transferred to Tabriz hospital where they stayed for a week. Their burns and blisters were treated with a white balm and they were given injections. Rebwar, however, who was more seriously injured, was taken to Tehran hospital where he stayed for a month. The survivors still suffer from skin rashes, chronic chest infections and seasonal health problems (see Chapter 4).

After this gas attack the inhabitants of Haware Khol did not dare to remain in the camp. Most people fled to the mountains nearby and found a resting place under boulders and in the valleys. About six weeks later, when Iran signed a peace treaty with Iraq they were relocated to Sera camp near Seqiz. It was further away from the Iraqi border and people felt safer there. Some people were still fearful in the new camp and they moved to live in the mountains nearby. The refugees in this camp were allowed to get out and to work. Not everyone had access to work, however, and many had no choice but to live on the minimal support provided by the Iranian government.

Refugees in Mardin camp in Turkey reported being poisoned through the bread they ate. Qadir[19] from Koreme village escaped with his family before the main road

---

19 Qadir, March 2010.

was blocked by the Iraqi army. He was one of the few men who was not executed in Koreme. The family ended up in Mardin camp where they stayed until 1991. One day in June 1989 when the refugees received their bread ration, hundreds of people fell ill. The symptoms included diarrhoea, stomach cramps, vomiting, slurred speech, disorientation, inability to walk in a straight line, and general weakness (Ala'aldeen et al. 1990). Turkey, however, suppressed the spread of this news and denied that the refugees were suffering from food poisoning. A Kurdish doctor in the camp who was the first to announce that the refugees had been poisoned was detained by the Turkish guards, according to Qadir, and was beaten and humiliated. Fortunately, blood and bread samples were secretly brought back to the UK by Jay Foran and Gwynne Roberts who visited the camp soon after. The examination pointed to 'a potent nerve agent (organophosphorus) as the cause of the poisoning' which affected more than 2,000 refugees including '667 children, 740 women, 663 men' (Ala'aldeen et al. 1990). No evidence was available as to who was responsible for this poisoning but judging from the context of Iraq's gassing the fleeing civilians and gassing refugees in Iran, it is likely that the Iraqi government was responsible for this attack. Fortunately despite the widespread ill health there were no casualties.

Migration from Turkey to Iran

After the September General Amnesty Turkey was encouraging people to return to Iraq but few people dared to. Some families gave themselves over to the Iraqi government, some stayed behind in Turkey and others headed to Iran. Asia's family, along with many others, were transported from their camp to the border of Iran. They were then picked up by Iranian cars and taken to Zewa where they stayed under tents for ten days in the cold weather. Then they were transported to Khoyee camp which consisted of buildings. Each flat was divided according to the number of rooms. Large families lived in single rooms. The toilets and kitchen were shared between two blocks. People had to queue to use the toilets. There were no washing facilities and everyone was allowed to use the public baths one day a week. The baths were open twice a week, one day for men and one for women. They were given basic food stuff which was enough to get by. The camp was surrounded by pasdars (Iranian security guards) and the refugees were not allowed to go out. 'It was a bit like a prison,' recalled Asia. Even when they needed a doctor the pasdars accompanied them to the hospital. After three months Asia's father started working and they moved to Razayeah and rented a small flat with other relatives. Within the space of a few months Asia's family moved 6 times (in Turkey and Iran). Their condition got better, however, as they moved from living under tents to living in proper accommodation in a city.

Another group of refugees who also headed to the border of Iran after the amnesty had to wait for three days before they were allowed entry.[20] Iran, aware of

---

20 Ahmad, February 2006.

its potential attraction for refugees in Turkey wanted to implement tougher rules to prevent a mass exodus. The Turks who had dumped the refugees on the border did not want to accept them back. After negotiations between the Kurdish parties and the Iranian government the refugees were finally allowed entry. They were taken to Zewa where they stayed for up to three months without access to tent or support. It was cold and people were hungry. Other refugees who had lived in the area long before Anfal helped the newcomers but the government refused to support them. Finally, the refugees were moved to Khoyee camp. According to Ahmad from Barchi village there were 1,750 Iraqi Kurdish families in the camp. This group of latecomers ended up staying for two years under awful circumstances. They lived in abject poverty. They were treated like prisoners and not allowed to leave the camp or to work. They had no access to private bathrooms or toilets, and their children were not allowed to go to school: 'We had fled one form of oppression to find another form of oppression'.[21] Finally, after another round of negotiations between the Kurdish parties and the Iranian government the refugees were allowed to leave. They went to Razayeeh where the men started working as farm labourers and they were able to rent proper accommodation. After the 1991 uprising the villagers returned to Iraqi Kurdistan.

Refugees' Social Experiences in Iran

As such large numbers of refugees arrived in Iran within days the camps could not cope with the influx. Some people were allowed to go to the towns and cities where they stayed first in the mosques and then in the homes of native Kurds who took them in and looked after them. Nazdar of Sireje,[22] who was eighteen years old at the time, nearly froze to death while escaping Kanitoo. The family, along with many others, were picked up from the border by Iranian trucks and they were taken to Bana. When she and her sister got off the truck, they could not stand on their feet and they fell, becoming the laughing stock of those who watched. They had spent three nights in the snow and their hands and feet were not functioning anymore. They spent two nights in a mosque in Bana, then a family took the two sisters home for a rest. Their hands and feet were blackened by frost bite and they could not wash, dress, or feed themselves. The woman they stayed with washed and changed them. She cut their hair which, because of the continuous freezing and unfreezing, was 'tangled up.'

A few days later the refugees were moved to Sardasht and the sisters ended up staying in a village outside town. While living in this village their nails fell off and their skin was peeled. Soon, Neworz (Kurdish and Iranian New Year on 21st March) was approaching and the sisters were concerned about their family. Their father was staying at a mosque in Sardasht. They did not know where their brother ended up. The rest of the family – their mother and the younger siblings, were

---

21 Ibid.
22 Nazdar, June 2010.

hiding somewhere in Iraq. On the eve of Nawroz, when the fires of celebration were lit, Nazdar and her sister started crying: 'In the past we celebrated Nawroz with our family. We used to prepare for Newroz and go to picnic,' Nazdar recalled. 'We were five brothers and three sisters living with our parents. Then suddenly we were separated into four groups. We cried so much thinking about these things.' The family tried to console the young women and tell them that at least they had survived but it was no use. They could not eat even though a feast had been prepared in their honour.

Migration entails geographical separation from the place of birth, a home, familiar environment, routines, friends and family (Veer 1998, Watters 2002). These losses may be more intensely felt at times of feast and celebrations when, under normal circumstances, the families get together and observe certain rituals. The losses can be so great that they impede a person's capacity to cope. They may leave the person in a state of depression (Stein 1986). Separation from the familiar environment and deprivation from the support and comfort of friends and family is accompanied by the challenges of adapting to a new world (Postero 1992, Veer 1998). Nazdar and her sister were eventually reunited with their father and brother and they hired a room in a village. It took a while for the rest of the family to join them. At first they were supported by the Iranian government and local Kurds. Gradually their father was able to find employment and the sisters managed to make friends and create new bonds. The process of normalisation took a long time and they were haunted by their traumatic experiences for about a year (see Chapter 6).

Women were forced to wear the hijab in Iran. Their families would not allow them to go out or mingle with the host community. Asia, who ended up living in Razayeah, pointed out that the young women were always cautious because they were being watched by their families and displaced community. They had no freedom, not even to go for a walk in the park because 'everyone was watching.' Most children ended up without schooling. Asia suggested that it was partly because they were poor and could not support the children's schooling, but it was also due to living in the poorer neighbourhoods with bad security. Parents thought it best to keep their children indoors, but there was no TV and no other entertainment. Children from the different families played together in the small space and sometimes they fought and argued, angering the adults. The refugee families were perceived as a nuisance by the host community. Some landlords would not rent flats to Iraqi Kurdish refugees because they had 'too many children.' Some of them were openly hostile and blamed the refugees for the inflation and unemployment. 'They blamed us for everything,' recalled Asia, 'they said that we had destroyed Iraq and now we were doing the same in Iran.' The landlord kept increasing the rent on Asia's family and he would not provide them with a landline. The house did not have a bathroom and all through their stay until 1991 the family had to use the public baths.

In summary, refugees were escaping grave dangers and they had embarked on long and exhausting journeys to reach safety. This safety, however, was an illusion for two reasons. First of all, despite crossing borders and surviving Iraqi assaults the civilians were not immune from continuing Iraqi attacks. The inhabitants of Haware Khol camp were gassed in July 1988 and the inhabitants of Mardin camp were poisoned in June 1989. These two incidents, even though isolated, were big enough to send a message to refugees in other camps: that they were not too far away from the Iraqi government's grip on their lives, that they had not escaped danger, and they were not safe. Secondly, the sense of danger continued in the bad security situation of refugee camp life. This was enhanced by hunger, illness, and lack of medical care.

Civilians found themselves facing a different range of insecurities and problems in exile. Some of them, especially those who had left Turkey to go to Iran after the September Amnesty, were given minimal support and not allowed to go out and work to better their situations. This group seemed the most vulnerable because of imposed helplessness and lack of crucial support for survival. Refugees, who ended up moving out of the camps, become entirely self-reliant. This reduced the helplessness that the refugee camp inhabitants experienced but it also made people face a various range of problems. In particular problems associated with finding appropriate housing and employment, facing xenophobia, living on minimal income, and dealing with isolation. The poor neighbourhoods they could afford to live in were not safe and they were not able to send their children to school. This was especially true for those who lived on the edge of large Persian cities and were deprived of social networks and kin relationships that are crucial for survival.

Internally Displaced Civilians

Some women during the Anfal operations managed to hide with relatives and acquaintances. The government attempted to capture them during the curfew with house by house searches and raids based on intelligence. The fugitives had to move regularly to avoid being found and to ease the burden on their hosts who could face severe consequences for harbouring Anfal escapees. This unsettled life, the constant fear of being caught, and the financial dependency played its role in making women feel helpless and depressed. Those who had many children reported telling their children off when they ate; they were ashamed of being a burden on the families who sheltered them.[23]

The situation of IDPs can be worse than refugees. Weiss (1999) clearly identifies the IDPs special vulnerability when he says: 'Those displaced within a country often are at least as vulnerable [as refugees], but they receive less attention and can call upon no special international agency, even though the General Assembly has called upon UNHCR to minister to all those in "refugee-like situations." Although the lot of refugees is hardly attractive, they may actually be better off than IDPs, whose existence customarily causes the issue of sovereignty to raise its ugly head.'

---

23   Ruqia, November 2005.

Similarly, Sehadri (2008: 31) points out that state sovereignty prevents aid delivery to IDPs and 'it is the task of the international community and global sovereignty as such to intervene on behalf of this vulnerable population and protect it from the scourges of civil war and ethnic conflict.'

This was particularly true for Anfal IDPs until the General Amnesty because alongside the losses and traumas experienced by refugees, they were still in imminent danger from the Iraqi government, they feared being reported by members of their own community, they felt that they were 'a burden' on their hosts, and they had no access to any support from any organisation. Roe (1992) describes the situation of women who are forced to flee their homes due to armed conflict, yet remain in settings of war and violence. These women, according to the author, experience terror, family instabilities, and sexual vulnerabilities. Anfal women IDPs, were completely powerless, terrified and at the mercy of those who sheltered them.

The Journey to Hiding

The majority of the women whose journeys have been described in this section have survived Anfal 3 of the Garmian region in April and the last woman, Nalia, managed to escape during the final Anfal in August. Rezan[24] from Salayee village was arrested and taken to Tuz Khurmatu Youth centre. A jash leader called Jabara Drej (Jabar the tall), who seemed to have been deceived about the purpose of this campaign, revolted against the state at the last moment and helped women escape. At the same time the people of Tuz Khurmatu demonstrated against the Iraqi government and the chaos helped some detainees flee. Members of the jash forces broke the window in one of the halls and helped women by using their kemmerbends as a rope to pull the women and their children out. On this day Rezan managed to escape with her two young children. The men pulled her children out first and landed them on the other side before rescuing her.

Rezan grabbed her children and ran into a nearby house of an unknown family. The family fed them and gave them fresh clothes. Their own clothes were muddy and filthy and this would have given them away. The head of the family then gave Rezan a lift to the neighbourhood where her sister-in-law lived. At first Rezan was scared because it was an Arab district and she was worried that they may report her. She did not come out of the house for three months. Soon Rezan found out that 'the Arabs were much nicer than the Kurds.' The wife of a senior Arab officer came and told them not to worry, 'We will not hand you over.' The Arab woman reassured Rezan and advised her not to move because she was safe there. She told Rezan that if she moves to a Kurdish district a jash leader may report her. This was true in many cases. Women who had managed to escape prison were reported by Kurds and then captured and taken away.

---

24   Rezan, November 2005.

Asmar[25] from Bangol village in the north of Garmian walked with her husband and 3 children in the pouring rain. The army was pursuing them and the region was being bombed. Helicopters were circling overhead. Older people were abandoned when they could not walk anymore, according to Asmar. Children lost their parents in the confusion. Everyone was desperate, shouting for help, swearing or crying: 'They talk about the judgement day,' recounted Asmar, 'that may be like Anfal.' The villagers had no choice but to head to Qader Karam. They were confused: 'No one knew what to do ... we didn't know what we were doing, no one was asking the other what they were doing... the people of the regions were all heading that way.'

On the main road to Qader Karam the jash and army arrested the villagers. They immediately separated the men from the women and that was the last time Asmar saw her husband. The families were then transported to Qader Karam where they stayed for two days before they were loaded into trucks once again. Everyone was hungry and exhausted by this stage and they did not put up any resistance. They were muddy and dirty and it was pouring with rain. Asmar got on one of the trucks with other women from her region. Unknown to her, her brother had come for her from Kirkuk. She saw him in the crowd and started crying. She was worried that he too may be arrested. Her brother kept urging her to get off the truck but she was paralysed with fear. He then started talking to the soldiers. In the chaos and pouring rain he managed to convince a soldier that his sister was not from the village. He insisted that she was only visiting when the attack started. The soldier who could not be bothered to argue, or maybe he felt sorry for them, allowed her off the truck: 'It was just luck. Think about it, he (her brother) stole us from the operation.'

Asmar's brother was well prepared. He had rented a car and carried a lot of cash. Asmar was praying all the way to the checkpoint. She was terrified of being recaptured. Her brother spoke fluent Arabic and told the Arab officers that he and his sister had been visiting a family in Qader Karam who had had a funeral. The General started swearing at him and saying that it was obvious that Asmar was a village woman. She was wearing Kurdish clothes which were all muddy and her skin was burnt under the sun. The officer said that he knew she was a farmer. Asmar's brother started pleading and begging. He kissed the officer's hand, which is a sign of inferiority and gratitude. 'I will never forget that day' Asmar said, 'my brother with his white beard reached out and kissed the hand of a young Arab General to protect us.' The man told them that he will let them off 'for God's sake.' Luck was with them and Asmar was taken back to Kirkuk. It rained all the way through the journey and Asmar was disorientated and confused. When they arrived her brother left her and the children in the car while he went home and changed. He did not take them home because he did not want anyone to find out that she had been rescued. That day they continued their journey to Baghdad.

---

25 Asmar, March 2006.

Asmar and her children were taken to an acquaintance's home in Baghdad where they stayed for five months until the amnesty.

Ruqia,[26] from Sey Jejni, had nine children when Anfal reached her region. Her eldest son and her eldest daughter were arrested and disappeared during Anfal. The rest managed to escape with her. In the cold and rain they hid in the hills for what seemed like two months. Members of the jash forces helped them avoid capture. During this long period of travelling from one village to the next and moving from one hideout to the next her baby daughter starved to death. Her other children were all hungry and ill. She was desperately trying to keep them quiet, in fear of attracting attention. Ruqia cried when she recalled how she had put pyjamas on her children's heads because there was no shelter and they were cold. Eventually a jash gave them a lift to Smood housing complex. Ruqia with her husband and children hid in Smood for a few weeks and then moved on to Kifri where they stayed in various homes until the amnesty.

Shamsa's[27] husband was taken to the Laylan animal pen where he stayed for about a month and then taken to be executed. She, along with her baby daughter ended up in Qader Karam where she stayed for two nights before the army started gathering people and transporting them. Every day between 9 in the morning and 2 in the afternoon trucks came to take the families to Topzawa. On the day when Shamsa was being shoved into a truck she ran from the soldiers' grasp. Many people tried to escape but most of them were recaptured, beaten bloody and forced onto the trucks. 'Some people broke their arms and legs trying to get off the vehicles,' Shamsa recalled.

In the chaos of shouting soldiers, desperate civilians, and crying children Shamsa quickly threw away the plastic bag which contained some of her baby's clothes and only kept her Iraqi ID card. She then picked up her daughter and ran. Her adrenaline was racing. She fell, got up once again and kept running without feeling fear or anger. The area was jam-packed with screaming and crying people who were either being loaded into trucks or were running away. The soldiers 'were beating women with the butt of their guns. They were beating them up in the trucks.' Shamsa ran and in the process she suffered a few blows from the soldiers. At the same time a few other people were trying to escape which distracted the soldiers and she managed to escape.

Shamsa ran into an unknown house. The family were eating lunch. She begged them to allow her to hide under their staircase for ten minutes. They asked her to join them for lunch. They had 'rice, broad beans, and meat.' They told her not to worry and not to be scared. They tried to comfort her but her whole body was shaking and she could not eat. The soldiers came into the courtyard pursuing her and she ran to hide under the staircase. The woman drew a curtain over Shamsa and her daughter, and told the soldiers that no one had come into their house. She told them they could search the house if they liked. Luckily they believed the

---

26   Ruqia, November 2005.
27   Shamsa, March 2006.

woman and left. The woman then apologised because she could not keep Shamsa and asked her to leave: 'She was worried that they would bulldoze her house over her head.' Shamsa left that house after resting for about an hour. She looked all around her before she took a step in any direction.

Shamsa had no one to turn to as her husband and her brother had both been captured. She walked in the opposite direction to the soldiers and sat by a wall. The sun was about to set. She spent the night in that road. Other people who had managed to escape were also lying around in the roads. There was no place to go to. None of the families from Qader Karam dared to give them refuge. They were hungry and some people came and gave them bread: 'Those who helped us will go to heaven,' Shamsa said, touching her heart. There was no place to sleep, no blanket or bed. She put her daughter under her dress to keep her warm that night. On the next day, she went into a ruin on the Qader Karam bridge. She stayed there for another four nights without shelter, food, or water, dependant on what people gave her to eat.

After spending about a week in Qader Karam Shamsa's husband's brother-in-law came for her. He lived in Kirkuk. She is convinced that if it wasn't for him she too would have ended up in a mass grave like many other women from her region. He paid 200 dinars at the checkpoint which was equivalent to $600 at the time. He too claimed they were from Kirkuk and had been visiting relatives in Qader Karam. He took Shamsa to Kirkuk where she stayed with a poor acquaintance for one month, and then she lived in various places.

Peri[28] from Goma Zard left her village in the night accompanied by her husband, two children and her extended family. They arrived at Qela Qochali road and her husband along with a few more men from the village decided to go back to the village to bring some goods. She waited for her husband in vain until she was told that her husband and the other men had been arrested. Then a member of the jash forces came and helped some of the families, about seventy people according to Peri, escape to Smood. They walked through the night. It was raining and people were up to their knees in thick mud. Some of those who tried to escape without help from the jash got lost and were eventually captured and trucked away.

After carrying both of her daughters for a few hours, around dawn, Peri was too exhausted. She kept the younger one who was a baby and left the toddler in a field: 'I just could not carry her anymore. I had no strength left,' she said. When the guide, the jash, noticed her he asked about the other child. She told him she could not carry her anymore and had left her behind. The man went back and found the three year old asleep in the field. He carried her on his back to safety. When they arrived in Smood in the morning the jash told them to be quiet otherwise they would get killed. The families that were willing to shelter the fugitives opened their door, others did not dare to give refuge to anyone. Peri ended up in a total stranger's house along with a few others: 'They helped us for God's sake and at

---

28   Peri, November 2005.

a cost to themselves.' The family sheltered them for a few days until their own relatives came to take them away.

During the final Anfal Nalia,[29] whose husband was shot in Koreme (see Chapter 2), managed to escape from Mangeshk school where she was detained with her two children and many other villagers from the region: 'Since my husband was killed, my father-in-law was killed, my brothers-in-law were killed ... I thought for whom should I stay in the prison? I prayed to God to help me break away from the grip of those oppressive people.' God answered her prayers, Nalia reported. The next day the prison guards turned a blind eye when she walked out of prison carrying her son and a sack of clothes. She had already given her younger son to her mother who visited earlier and she gave her older son to a relative in Mangeshk. She then walked into the house of a total stranger, gave them the sack of clothes and begged them to give her a bucket. She carried the empty bucket on her shoulder and pretended that she was a woman from Mangeshk, going out to bring water. She walked to her uncle's house in Mangeshk. From there she got an abba (a cloak women wear to cover up from top to toe) and she went to the garage where a man she knew promised to give her a lift should she never mention his name. He gave her a lift to Duhok and then Nalia was helped by relatives to reach Mosul.

Life in Hiding

> People were scared of their neighbours then. You were always scared of being betrayed. There was no trust. (Peri, November 2005)

When women went into hiding during Anfal they became financially dependent on relatives and acquaintances. They were also housebound and had to keep their children indoors for months. The main issues for these women were the fear of being caught, guilt for becoming a financial burden, frustration and anger when all they could do was to wait, and sadness and guilt over their children. Rezan survived two raids by pure luck. One morning, when she was living in a room in her sister-in-law's courtyard, there was a violent bang at the door. This was an un-announced curfew. She was having breakfast with her son in her lap. Her sister in law made a gesture to be quiet and she put a lock on the door from outside. When the security forces came in the woman told them that this part of the house was empty for rent. Rezan was 'terrified' inside, she started breastfeeding her son to keep him quiet and she held on to her daughter. To her surprise the security officers did not ask her sister-in-law to unlock the door. They did not search this part of the house.

Many people, who were less fortunate, were arrested and taken to the Dibs prison on that day. Rezan stayed in this house for three months then she rented a small flat in a courtyard in Tuz Khurmatu. Soon after she moved in there was another raid and the landlord similarly put a large lock on the place saying that

---

29   Nalia, April 2009.

it had not been rented. The neighbors, many of whom were Arabs, testified that the flat had been empty for a few months. Rezan heard the soldiers anticipating in the road outside. This time they were members of the Republican Guard and they spoke in Arabic. Rezan is grateful to the Arab families who did not report her. This is particularly because she expected the opposite of them. She is also disillusioned with her own community and in particular the jash who deceived people to surrender and reported others during the curfews: 'They did the worst things for money,' Rezan recounted. She is also disappointed with some of her own relatives who told her: 'Please don't come to us, we don't want to be arrested.' Rezan does not speak with them anymore because 'at the time we needed them the most they did not help us.' She still remembers all of that and she remembers all the good people who risked their own lives to protect total strangers.

After rescuing Asmar from Qader Karam her brother took her to the house of an acquaintance in Baghdad. She was accompanied by three children and her last one, a baby boy, had severe disability. When she was eight months pregnant Asmar had a major shock when a plane dropped its bombs too close to her. After that her baby did not move for three days. She believes her son must have suffered from a stroke when he was inside her. The family Asmar stayed with were Arabic and the man had worked with Asmar's brother in the past. She stayed with them for over five months. Her brother used to send her money. She was in Qadisiya, a neighbourhood named after Iraq's war with Iran. She did not speak any Arabic and was stunned by the vastness of the city: 'We were villagers, we weren't used to the large place.' Every time the family had visitors people enquired about who she was. They told people that she was a relative but Asmar was always scared of being informed on. She was embarrassed about bothering this family. Her children who were not used to being confined in a room were 'crying all the time.' She had no friends and no acquaintances there and she did not have any news of her husband or other relatives. Totally isolated and lonely, she cried regularly when she was alone with her children.

Sometimes Asmar wished she had been in prison instead. 'It would have been better,' Asmar said but she did not want to expand on the reason. Once after two months of staying in Baghdad she went back to Kirkuk because a relative had passed away. She missed her family and she used the wake as an excuse to return to Kirkuk. She had just arrived at her parents' home and her sister was about to serve them lunch when the security arrived at the door. The officer said that they have been informed that a villager called Asmar is staying there. Somebody had informed the security forces of Asmar's return. She managed to escape from the roof (the roofs in this part of the world are flat and connected). She walked to the neighbour's roof and they let her out from the back door. She then returned to Baghdad and stayed put until the amnesty.

Ruqia, with her six children, initially lived in Smood. She hid in her brother's house who welcomed her and said, 'I will protect you. Let them take me with you if they come.' They closed the door on Ruqia and her children. But there were regular raids in Smood and she decided to move. She stayed with a different family every few days- her brothers, her sisters, her cousins: 'We were like cattle, being moved all the time,' she recalled while crying. Some families were not welcoming but Ruqia had no choice but to bite her lips, feel embarrassed, and stay. She waited impatiently for everything to end. She also remembers how some families, who were themselves poor and scared, gave them food 'grudgingly.' She kept telling her children off and asked them to eat less: 'We locked the door on our children and didn't let them see the sunlight.' Ruqia now cries remembering those days. She feels sorry for her children who were young then. Some of them don't even remember any of this but she cannot forget the injustices they suffered and how she was harsh with them. She recounted with great sadness that her children did not have enough food, they could not come out and play in the courtyard. They had no toys, and they could not go to school.

Shamsa was brought back to Kirkuk by her husband's brother-in-law. He did not dare to take her back to his own house or the house of any other relative. She was taken to an acquaintance's house. 'From then on', Shamsa said, 'we were like beggars, each day in a different house.' Shamsa and her daughter spent about five weeks with this family in Kirkuk who had one room. They were 15 people already and they had a disabled son. Everyone slept in that same room, 'on top of each other.' The host was a really good person, very generous and big-hearted. He kept saying that he would put her 'on his eyes' – an expression meaning you are very welcome- but Shamsa felt embarrassed. The family were already overcrowded and the man himself was a poor labourer who had enough mouths to feed. After that Shamsa returned to Chamchamal and moved in with her parents-in-law who were ill and bedridden (see Chapter 5, section on accommodation). Unlike many village women Shamsa, who was originally from Kirkuk, had an Iraqi Identity Card. Most people did not know her in Chamchamal and they did not know that she was supposed to be arrested. She was still scared, however, because the government kept doing searches and they managed to capture many people who had fled the Anfal campaign.

Nalia stayed at her uncle's house in Mosul. She was accompanied by her baby son and she did not know about her older son who was being moved from house to house to hide him. She herself moved four or five times in the one month she was hiding. Each family she stayed with including her father, her uncle, and her cousin's family, were all scared of being found out. She also did not want to cause them trouble. One night she heard her uncle talk to her father on the phone. He had been warned that he and his sons would be killed if they harbour a fugitive. Nalia started shaking and crying upon hearing this. She wanted to leave her uncle's house that night because she was scared for him: 'He told me that he wouldn't let me go even if they shoot him and his two sons that night. He asked me where would I go? I said I would go stay behind the walls until the day breaks. He said let them kill me, I won't let you leave this house tonight.'

Nalia had been two months pregnant when she was separated from her husband and she heard the shooting that killed him in the village. This trauma combined with the bumpy journey in the military truck and the fear that followed her after her escape caused a miscarriage. When she started bleeding in Mosul the doctor told her that she needed complete bed-rest if she was going to give the foetus a chance. However there was no one to look after her 11 months old son who suffered from severe diarrhoea. She made a rational decision not to try to save the pregnancy: 'I thought I have raised this one (the toddler) and he is a bit older, I don't know what will happen to the other one (the foetus).' A few days later she lost the baby but at least her toddler survived.

The dominant themes in the lives of internally displaced women during Anfal were exhaustion, stress, fear, guilt, claustrophobia, and helplessness. After the nerve-racking journeys they had undertaken the women were unprepared for confinement in closed rooms with their children, the constant state of anxiety, the feeling of being a burden and their inability to be good mothers to their children by not being able to give them enough food and not letting them go out. They were also disappointed with those friends and families who failed to help them at moments of need. This rupture of bonds was crucial to make them feel alienated and isolated from the rest of the community. On the other hand help provided by unexpected others had a positive effect in restoring faith in humanity.

Conclusion

Migration entails dealing with loss and powerlessness, as well as adaptation to the new environment. The losses include material and psychological things such as home, familiar environment, way of life, support networks, culture, and employment (Waters 2002, Espin 1992). During their resettlement refugees experience what Veer (1998: 22) calls 'a sequence of displacements' when they are repeatedly moved from one place to the next without being consulted. This usually leads to feeling powerless and being at the mercy of the host country. This is very similar to what Hiroto and Seligman (1975) named 'Learned Helplessness.' These authors found that when people learn that they have no control over unpleasant events they become helplessness and this causes depression. This was confirmed by Lengua and Stormshak (2000) who found that an external locus of control (having no control over unpleasant events) predicts more avoidant strategies and higher levels of depression and antisocial behaviour. However, Harrell-Bond (1986, cited in Baron et al. 2003: 248) suggests that this 'dependency syndrome' can be seen as a result of the way in which aid is distributed and managed in refugees settings. Baron et al. (2003: 262) suggest that aid efforts which involve giving some responsibility back to refugees and empowering them by promoting food for work schemes and encouraging farming, are generally more effective than others that merely distribute food and medication.

The situation of Anfal IDPs was different from many others in the world. First of all, they were not living in camps or settlements like other IDPs (Kim et al. 2007, Roberts et al. 2009) but were scattered in the community, isolated, powerless, and dependant on the generosity and kindness of their hosts. Secondly, no international aid agencies had access to them. They were a hidden and invisible group which were not supported by any national or international organisation. Thirdly, they were on the run from the same government which was supposed to protect them and because of this they had no fixed location and had to keep moving every few days or weeks to avoid being arrested. This state of being in a flux in itself was unsettling and it prevented the reformation of support network or routine.

Dangerous and risk taking behaviour, like trying to escape against all odds (Shamsa and Nalia) helped some women survive. The majority of those who escaped, however, were helped by various individuals. The help that saved many lives could be as small as giving a woman a fresh set of clothes (Rezan) or a bucket (Nalia) so that she passed for a town woman, giving a fleeing woman shelter for a few minutes (Shamsa), or a lift to the nearest safe place (Rezan and Nalia). More risk-taking help involved guiding a woman into safety (Peri), negotiating with guards and paying bribes at the checkpoints (Asmar), and giving refuge to women relatives and acquaintances who were on the run. The women who were helped are forever grateful for the aid they received, especially for the help provided by individuals who were perceived to be on the enemy's side (members of the jash forces and Arab families). On the other hand, betrayals by others who were expected to help (family, relatives, and the Kurds in general) are experienced with great pain. It has made women feel even more isolated, desperate, and angry.

Women, who had no choice but to restrict their children, tell them off if they made noise or 'ate a lot' now feel guilty and sad. While hiding they were not capable of taking any active measure to improve their situation. They could not work to earn money which would have made them less dependent on their hosts. It is also possible that some of these women were abused by their hosts and relatives but this was never talked about by anyone. Life got better after the amnesty when women could rely on themselves more and try to provide a home for their children (see Chapter 5). Until then, however, all they could do was wait for the situation to change.

There seem to be various individual strategies to survive migration and internal displacement. Within the constraints imposed on them by their situations people tried to change things for the better. As refugees they moved from one camp to the next, from the camps to the cities, from one country to the next (Turkey to Iran) searching for a better life. The internally displaced also moved from one hideout to the next hoping to ease the burden on their hosts, to have more control, and find a more secure and suitable location. Sometimes survival was random and outside the individual's control. Freezing to death, falling off a mountain during flight and falling ill without access to medication in the camp, being gassed, or poisoned were unlucky events that could not be prevented. Similarly, being picked up during a raid, or being informed on, and not being helped were events that the internally displaced had no control over. Luck seemed to play a role in survival as did taking risks, taking control, and help by others.

Chapter 4
Survivors of the Gas Attacks

We are as good as dead. We are ill, we cannot work, we have no power over anything. (Fatma, April 2006)

Iraq repeatedly used chemical weapons in its war with Iran. The first widely reported gas attack was during the summer of 1983 in the Iraqi Kurdish town of Haj Omaran, near the border of Iran. This was in retaliation for the July-August Iranian offensive during which KDP peshmarga led the Iranian army to the area (Hiltermann 2007: xiv). The regular use of gas cast terror in the heart of Iranian soldiers and it famously led to the decline in the enlistment of young volunteers. Throughout the 1980s Iraq showed a TV program called 'Images from the Battlefield- Swar min al-maaraka.' Some of these images showed, amongst other examples of violent death and disfigurement, dead Iranian soldiers whose faces were blackened. At the time there were rumours in Iraqi Kurdistan that these were victims of gas attacks but no one knew whether these whispers were true.

In April 1987 the Iraqi government, for the first time, used chemical weapons against Kurdish civilians in the Jafati and Balisan valleys. This was in response to the PUK's attack on Iraqi military posts in the Jafati valley (MEA 1993: 60). On 15 April 1987 Sargallu, Bargallu, Haladin, Kanitoo, and other villages nearby were attacked with chemical artillery shells. This attack caused few civilian causalities which was not the case with the next attack on the following day. On 16 April 1987, Balisan, Shekh Wasan, Malakan, and Tootma were gassed by a dozen airplanes. Even though the region was a local PUK headquarters there was low peshmarga presence at the time because the majority had been called away to fight in Jafati valley. The extensive gassing targeted large civilian populations in the Balisan valley. According to research carried out by Middle East Watch between 225 to 400 civilians died as a result of this attack (MEW 1993: 70). To this number must be added those who died in the following months and years as a result of chronic health problems related to the chemicals.

Chemical weapons were used at the beginning of every Anfal stage throughout February to September 1988. Alongside killing and injuring the attacks served to terrify and uproot civilians from their villages which, in most cases, lead to their arrest and then murder (see Chapter 1). During the first Anfal of the Jafati valley (23 February-19 March) the PUK tried to distract the army from the region by opening another front. On 15 March 1988 the peshmarga along with Iranian Revolutionary Guards seized Halabja, a town of about 80,000 on the border of Iran. On the next day Halabja was heavily gassed.[1] These attacks lead to the death

---

1   For a detailed account of what happened in Halabja see Hiltermann (2007).

of thousands of civilians on the spot. More died during the exodus to Iran and in
Iranian camps and hospitals. In the following years other people lost their lives as a
result of developing long term health problems associated with the gas exposure.

At first, it was unclear what type of gas was used by the Iraqi government.
After a nine day visit to two small refugee camps in Turkey (7-16 October 1988)
Physicians for Human Rights concluded that the blistering that many survivors
sustained was consistent with 'chemical burns by a blistering poison gas such
as mustard gas' (PHR 1989: 1). Some victims, however, died within minutes of
exposure to the gas which could not be explained by mustard gas alone. PHR then
concluded that at least one other nerve agent had been used in addition to mustard
gas (PHR 1989: 2).

Iraq used a combination of gasses and nerve agents which is regularly referred
to as 'a cocktail of deadly gasses' in the literature. However, there was disagreement
as to which nerve agents had been used. Heyndrickx (1988), who visited Halabja
few weeks after the gas attack and collected samples, concluded that mustard gas
and cyanide as well as nerve agents such as tabun, sarin and soman were used. The
use of cyanide was disputed, however, since Iraq was not credited with possession
of this chemical (Hiltermann 2007: 193-196). Some, including Iraqi officers and
Iranian doctors, also believed that VX was used but this has been disputed by
others.[2] It is now agreed that the cocktail of gasses included mustard gas as well as
nerve agents such as tabun and sarin and possibly others.

According to the research carried out by the Halabja Post Graduate Medical
Institute 250 inhabited areas (towns and villages) as well as 31 uninhabited strategic
areas in Iraqi Kurdistan have been gassed by the Iraqi government between 1987
and 1988 (Baban: 2000). The effects of exposure to the gas varied according to the
person's distance from the bomb site, the wind direction, whether they had access
to wet towels, and the type of chemicals used. Those who were within meters of
the bomb and were exposed to nerve agents died immediately. Others managed
to survive and some were initially unaware that they had been injured. The gas
attacks usually made a muffled sound which at times made civilians think that the
explosions were far away. Then clouds of yellow, brown, or pinkish smoke spread
over the region. The gas reportedly smelt of garlic and apples or, in some cases,
pesticides. Soon after inhaling it people started suffering.

The different chemicals produced different symptoms. Mustard gas exposure
led to blistering, tightness of breath, and burning throat and eyes. There is no
antidote to mustard gas and the only way of avoiding long term damage is burn
care, rapid decontamination of body and clothes, and eye therapy (Gosden and
Gardener 2005). The nerve agents resulted in temporary blindness, blurred vision,
tightness of chest, runny nose, coughing, nausea and vomiting, increased sweating,
salivation and tears, fatigues and muscle paralysis, loss of consciousness and coma
(Asukai et al. 2002: 150). Atropine injection can be used as an antidote to nerve
agents but the majority of civilians did not have access to this. People had been

---

2   Hiltermann (2007: 142-143).

advised by the peshmarga to build a fire and wrap wet towels around their faces. In most cases, however, the sudden attacks caught people unprepared and hardly anyone had time to light a fire. Some managed to use wet towels but many people who inhaled the gas, those who were exposed to the gas through wind, and others who tried to reach the springs and rivers to wash the gas off their bodies died.

The helplessness felt in the face of chemical weapons was unlike any other. By 1988, most people were used to conventional bombing. Civilians had learnt to protect themselves by hiding in shelters and caves and the peshmarga were able to either hide to avoid injury, or fight back. However, when it came to chemical attacks everyone was at a loss. 'How can you fight chemicals?' recounted Adiba[3] from Derkari Ajam in Badinan, 'they were in the air we breathed.'

Beaton and Murphy (2002) point out that even the threat of a Biological and Chemical Weapons (BCW) attack can cause major anxiety and health problems. This is because people are aware that these weapons can cause large scale civilian casualties. They are unpredictable in terms of how they affect the victims' health in the long term, are difficult to detect, and, unlike natural disasters, they are the result of intentional and malicious actions. The authors then go on to conclude that: 'The potential for BCW agents to instil fear in the general populace is magnified by their ability to inflict harm without being readily detectable. There is the potential for panic, social unrest and economic dislocation in the aftermath of an actual BCW attack.'

Little research has been carried out in Iraqi Kurdistan about the long term health consequences of chemical weapons for the exposed population. Christine Gosden's research (see Gosden 1998a, 1998b, 2002) whilst anecdotally illuminating, fails to provide a scientific basis for her conclusion that the widespread cancer, infertility, and chronic ill health she observed in Halabja were due to the gas exposure. Because of a lack of reliable evidence, I will concentrate on the survivors' experiences of the gas attacks and the aftermath. Whether or not some of the reported illnesses are related to gas exposure is beside the point. What matters here is how women see themselves as continuous victims of the gas through health problems, community stigmatisation, and the constant fear that they report about their health and future prospects.

The Gas Attacks

In the following sections I will outline the situation during some of the major gas attacks on villages between 1987 and 1988. Halabja has been excluded from this account for two reasons. First of all, this book is about Anfal which targeted villages declared 'prohibited' by the Iraqi government in the 1980s. Halabja was a town and was not categorised as 'prohibited'. It was not part of the Anfal campaign. Secondly, the story of Halabja is a complicated and lengthy tale which deserves a

---

3   Adiba, February 2006.

book in itself which is not in the scope of this research.[4] Here I will focus on the situation of survivors of the following major gas attacks: The April 1987 gassing of Balisan valley, the February 1988 attacks on Jafati valley (Anfal 1), the March 1988 gassing of of Sewsenan (Anfal 2), the May 1988 gassing of Goptapa (Anfal 3), and the August 1988 gassing of Badinan region (final Anfal).

Balisan Valley, April 1987

On 16 April 1987, the inhabitants of Shekh Wasan and Balisan villages had just arrived home after a day's work in the fields when twelve planes appeared in the sky. This was nothing new. The valley had been declared 'prohibited' since 1983 and it was bombarded regularly throughout those years (MEW 1993: 61). There was one year of peace when the PUK was negotiating with the Iraqi government (1985) after which the bombardments intensified.

Mehabad[5] from Shekh Wasan village was at home with her two children, her brother and his family, and her father-in-law when the planes dropped their load in the village. Her husband, who was a peshmarga, was on duty in the Jaffai valley, taking part in an offensive against the government.

The impacts were not loud which ironically made the villagers feel safe. According to Mehabad an old man shouted to inform the people that the losses were only one person and some poultry. The civilians were reassured that despite the involvement of twelve planes, this time they had got off lightly. Mehabad stayed at home while a few helicopters followed the planes to check on the village. She still did not realise what was going on and she told her neighbour, another woman whose husband was fighting in Jafati valley, that may be it would be best to leave. She was worried that the helicopters may drop soldiers into the village and the two of them would be vulnerable because their husbands were peshmarga. However, her neighbour believed that it was late in the day and it would be best to leave in the morning. Mehabad went back inside the house and the family started feeling rough in the hours that followed. Their skin blistered (an indication that mustard gas had been used) and they threw up (due to nerve agents). They were all temporarily blinded and their bodies were hurting and swelling.

Similarly, Sabiha[6] from Balisan village took shelter when the planes came. She went in along with her seven children and some other villagers. She recounted that when they came out 'white dust' landed on their clothes and it started raining. The chemicals drenched into their clothes and started burning their 'skin and eyes.' Soon everyone in the village was desperate, all screaming from pain and weeping. No one was able to help the other. In the morning inhabitants of the nearby villages guided the villagers from Balisan and Shekh Wasan to the mountains. They spent one night in the caves and mountains, their noses and eyes streaming, their bodies

---

4　For a brilliant account of the events in Halabja read Hiltermann (2007).
5　Mehabad, April 2006.
6　Sabiha, April 2006.

blistered, and all of them blinded. On the next day tractor-drawn carts were brought to take the people to Rania, a small town nearby. Mehabad is not certain who organised this but according to Sabiha it was the jash who arranged their transportation to Rania where there was a hospital and they could get treatment.

In Rania civilians looked after the wounded villagers and local doctors gave them eye drops and washed their wounds. The next morning security agents arrived and took the wounded villagers to Erbil. Initially they were taken to Erbil hospital but before Mehabad's and Sabiha's families received treatment (both women and their families waited their turn in the trucks as others were tended to in the hospital) the security agents took them to the security office in Erbil. The majority of the villagers, according to Mehabad and Sabiha, were blind and they were 'weeping' and praying to God for mercy.

Mehabad kept stretching her hands and checking that her two children were still with her. Her four year old son was suffering particularly. He could barely cry. On the third day a woman advised her to go out and wash her face because this may help her feel better. She went to the prison courtyard, guided by a woman, but when she came back her son's place was empty. He had been taken by the guards: 'He was still alive when they took him,' she said while crying. Her young niece also passed away in prison. Mehabad considers her son a victim of Anfal although Anfal started 10 months later. The gas attack on the Balisan valley is directly linked with Anfal for most of the villagers. It was a warning as to what was to come. By launching this attack Iraq tried out the potential of its new weapon on civilians and it also tested the world's reaction to gassing civilians. In the absence of any reaction the planning for Anfal went ahead (Hiltermann 2007: 16-17).

The families spent about one week in prison, sleeping on the concrete floor without food, medication, or cover. Then one morning the men were separated from their families. They were taken in buses and were never to be seen again. Mehabad's brother and her nephew were among the men who were taken on that day. Sabiha's husband, her eldest son, two of her brothers, and her father were also amongst the group. Sabiha's husband managed to speak a few words to her before he was taken away. He asked her, 'Have your sons died on you?' She replied that they have not. He told her, 'Be careful, don't let them die.' And then he was gone.

The women and their children were then dumped near Khalifan, a small town three hours north of Erbil. They were helped by Khalifan civilians, given new clothes, washed and fed. On the next day Mehabad's surviving brother who lived in Shaqlawa came and loaded them into the boot of his car. He was scared to take them in the open. Mehabad and her daughter stayed at her brother's house for two weeks where the doctors came to treat them secretly. They did not dare to go to hospital in fear of being arrested once again. She then reunited with her husband in Shekh Wasan and lived there until Anfal came to their region a year later. Fortunately the family managed to sneak back into Shaqlawa and avoid being captured during Anfal. Sabiha and her six remaining children also ended up in Shaqlawa with her brother-in-law's family and remained there till the early

1990s when Balisan was rebuilt and then they returned to their village. The events mentioned above are consistent with the description provided by Middle East Watch which was also based on eye witness accounts (MEW 1993: 60-73).

Jafati Valley, February 1988

Between 25 and 28 February 1988 a cluster of villages were gassed at the beginning of the first Anfal. The assault targeted Yakhsamar, Sargallu, Bargallu, Haladin, Chokhmakh, Gwezeela, Chalawa and surrounding mountains.[7] This caused a mass exodus towards Iran (see Chapter 3). Khucha[8] from Chalawa, had already left the village and lived, with her husband and brothers-in-law, in a small house on Genderan mountain. A friend of her husband came to visit one day.[9] He too had his family with him and he asked Khucha to tell her husband not to leave him behind when they decide to flee. He left and at that moment there was 'a quiet explosion nearby,' Khucha recalled. She quickly got up to shut the door in order to protect herself from possible shrapnel. The force of the impact, however, lifted her into the air and threw her against the wall. As she was losing consciousness she saw 'a pink dust' entering the room.

When Khucha woke up her husband was spraying water on her face. His friend had died instantly. Her husband too had been affected by the gas but he seemed to be better than her. Khucha's eyesight was blurred and her body was heavy. Her husband carried her on his back to Zindanan Mountain where a few peshmarga gave them atropine injections against the nerve agents. The peshmarga who tended to them, however, started suffering as they inhaled the chemicals from Khucha's and her husband's clothes. This is called 'secondary contamination' when individuals exposed to people and areas that have not been decontaminated become exposed to the chemicals and suffer as a result (Beaton and Murphy 2002).

After resting with the peshmarga for a day, Khucha and her husband walked up the mountain on all fours to get to the cave at the top. They had two sacks of bread with them. Her husband suggested that they should offer the bread to the donkey before they themselves eat it. The donkey smelt the bread a couple of times and walked away from it. They threw it away because they now realised it was contaminated. Snow was melting and streams were coming down the cave's walls and ceiling. Khucha and her husband went under the cold streams to wash away the chemicals. On the next day they started their journey towards the border of Iran supported by people who held them up and helped them walk.

In Galala the couple met Iranian soldiers who gave them eye drops and put them, along with a few other injured villagers, into a tractor-pulled cart to Iran. After a short time the tractor broke down and the passengers spent the night under snow and rain without food or water. Everyone was 'blind and lame,' according to

---

7   Based on research carried out by The Ministry of Anfal and Martyrs Affairs, 2007.

8   Khucha, April 2006.

9   Like most villagers Khucha is uncertain about exact dates.

Khucha. On the next day a car passed by and took them to Bana in Iran. Khucha was taken to the town mosque. She had been gassed about a week before and had spent a few nights in the rain and cold. When two women held her arms and took her into the heated room in the mosque, she fainted. They then wrapped her in a blanket and took her to the first floor where there was a pool. They had sterilised the pool and washed her three times a day which may have assisted decontamination, even when quite late.

Khucha recounted not being able to stay in a warm room because the heat made her feel worse. After being looked after in the mosque for a few days she was taken in by a Kurdish family from Bana. The woman washed her three times a day, changed her and fed her till she was a little stronger. Finally she and her husband were transferred to a refugee camp near Bana. Four months later, in July 1988, Hawara Khol camp was gassed by the Iraqi planes and Khucha was wounded for the second time (see Chapter 3). Her eyesight and skin were affected once again and this time she was taken to Seqiz hospital where she was given drops and injections. Khucha stayed in Iranian refugee camps until the amnesty was announced in September 1988. She had the doctors' reports and her medications but she was advised to get rid of them before she reached home: 'They told me Saddam will hang me if I keep the documents.' At the time any talk of chemical attacks could lead to people being taken by the security agents and then disappearing. Khucha finally returned to Chalawa in the early nineties when the village was rebuilt.

Qaradagh, March 1988

On 22 March 1988 the second Anfal attack reached Qaradagh region with a major gas attack on the village of Sewsenan. It was the second day of the year according to the Kurdish calendar. Bakr[10] was having dinner with his large family – his wife and five children, his mother, brother, sister, and a guest. The village was heavily bombarded by truck mounted multiple-barrel artillery. The muffled sound made Bakr think that they were bombing Belekjar, the next village where the PUK Regional Command was based. His young son, Hemin, was really scared but Bakr assured him that it was nothing serious.

Bakr's neighbour came to hide in their shelter with his wife and children as they did not have a shelter of their own. Unknown to them a shell had landed in Bakr's garden, near the shelter. The neighbour's wife screamed that her husband was dead.[11] Bakr thought that he must have been injured by shrapnel, he walked out to check on his neighbour and then quickly returned inside to his family. He wasn't sure what was going on. The family and their guest headed towards the shelter and Bakr went to fetch some wet towels just in case. He never made it back out as he immediately fainted in the room. Later he was picked up by other villagers from

---

10   Bakr, April 2006.
11   Later Bakr found out that his neighbour himself had survived while his entire family perished.

the region. He was the only survivor in that family. His wife and five children along with his mother, sister, brother and guest as well as the neighbours had all died on the stairs to the cellar and in the garden. His three month old daughter was also in the same room with him. She had died in her cradle.

Civilians took Bakr to the PUK hospital where he was given injections and treatment. He was unconscious for a few days and many people thought that he would not make it. The commander ordered the peshmarga to leave him behind because he was 'as good as a corpse,' Bakr was later informed. However, two of his cousins carried him in a coffin over the Darmazalawa Mountain. Other relatives from Suleimanya who had heard about the gas attack came and managed to sneak him back to Suleimanya in a car. He was hiding in his cousin's house and was secretly being treated by a good doctor who checked on him on daily basis. His face, neck and chest were all blistered. He could not breathe. He was blind and unconscious most of the time.

For six weeks Bakr was told that his wife and children had survived and that they were hiding in Derbendikhan. Only when he was getting stronger, when he could see and walk again they told him the truth. His relatives came together and his elder cousin told him: 'People don't suffer blows every day' and 'a man should be strong.' He was informed that all members of his family had died. They then took him to Bakrajo, a district at the edge of Suleimanya. They left him walking up and down the fields of Bakrajo all night while his inside 'was burning like an oven.'

The victims of the Sewsenan gas attack are now buried in the Martyrs graveyard. There are 85 graves of people who were buried in their own clothing after the gassing. Their bodies were recovered and gathered in a special graveyard in 1993. There is a statue of a gas shell in the middle of the graveyard to remember the cause of their death.

The Valley of the Lesser Zab, May 1988

The fourth Anfal started with major gas attacks on the villages of Goptapa and Askar on 3 May 1988. Goptapa was a wealthy and large village consisting of 452 extended families (Resool 1990: 96). Located next to the lesser Zab, the village was a perfect location for farming and animal husbandry. By 1988, and because of the regular bombardments, most of the villagers lived in shelters and temporary accommodation outside the village. Some, however, had remained in and around the village despite the rumours of forthcoming government attacks. In the afternoon of 3 May, women were preparing dinner and men were heading home from work when the planes came to drop their load.

Aram,[12] who was 16 years old at the time, was in his shop in the east of the village. He saw 'the large planes' that circled the village but he was not alarmed. 'These big planes, Sukhhoi'[13] he recounted 'didn't fire at people, not like the other

---

12   Aram, February 2006.
13   Soviet made planes.

planes. Then one of the planes threw a balloon into the air. It was to test the wind. It measured the air direction and movement. Then they came back. I think it was six or eight of them that returned to the village and started gassing.' The bombs were widely spread, 'so that the largest number of people would be killed,' Aram argued. The wind brought the gas in their direction and everyone started fleeing to the south, away from the gas bombs which targeted the middle and north of the village.

The men ran through the southern part of Goptapa which was empty. It was the distance of one to two hundred metres which Aram believes was long enough to take one breath. They held on to their noses and mouths as they ran. Helicopters then came to check on the village and the men were worried that they may be shot. They first hid under the bridge near the hospital and then they went to the top of a hill because they had heard that heights protect from the gas. Aram watched as the wounded civilians came to the hill, running for the pond, created by rain water. Their bodies 'were burnt by the gas' and they were 'screaming from pain.' A doctor and some peshmarga went into the village and came back carrying injured children. The pehsmarga kept bringing the injured and giving them atropine injections but most of the people who were brought there died.

Two hours later, at about 8 in the evening, Aram went back to seach for his family. Nobody was at home or in the shelter and he hoped that this meant they had escaped. He then walked towards the stream behind their house and found that every 5 to 6 metres another member of his family had fallen facing down, in an attempt to get to the water. 18 members of his family died in the Goptapa gas attack that day including his grandmother, mother, four sisters, brother, uncle, aunt, and his cousins. Their corpses littered the area. Some of them had a towel in their hands, hoping to wet it in the river but they had not managed. Some, including his sister, had a gas mask but they had not known how to put them on. Two of his girl cousins had embraced each other and died in each other's arms.

Aram was largely disorientated. He felt confused. His body was heavy and his eyes were burning. 'We had no means to do anything,' Aram recalled with despair, 'we were completely powerless.' He did not find his mother's corpse and he was encouraged by the others to leave. Unknown to him his mother had fled to the broad beans field, carrying the cradle of his baby sister. She had fallen and died there. The cradle had fallen into the field. On the next day the jash who passed through the village, looking for loot, found the baby crying in her cradle. One of them picked her up and took her to hospital in Chamchamal. It took about a month before the family was able to trace her and get her back.

Now 105 victims of the attack on Goptapa have been gathered in a cemetery on the green hill. In the middle of the graveyard there is a statue of a screaming old man, outraged by the catastrophe. Not all the victims gathered here are from Goptapa,[14] some of them are nameless peshmarga and deserters who were in the village or

---

14 Muhamad, February 2006.

surrounding areas at the time. This is another simple monument put together by survivors to remember the victims and seek closure (see also Chapter 6).

Badinan, August 1988

During the final Anfal more than a dozen villages were gassed and this caused massive panic amongst the civilians and made them flee towards the border of Turkey. Aysha[15] from the village of Bandaw was fourteen years old when Anfal reached her region. The family were already living in a cave because their village was burnt down two years earlier. With her extended family Aysha started a march to the border. During the day they hid in gorges, under boulders, and in forests and they walked through the night. Twice they tried to cross the main road to Turkey but all the routes were blocked. Aysha, who carried her nephew, was exhausted and hungry. Her brother, who himself was carrying two children, told her to leave his son and keep walking. But she refused to abandon the boy.

On the third day the group turned back to the gorge of Baz and Belizan. This is a fertile valley with plenty of tall trees. They gathered in the valley to rest and they were joined by hundreds of others: 'The whole world was there,' recalled Aysha. A man who had gone up the mountain to keep watch ordered people to jump into the river as there was a gas attack. He had witnessed the gassing of a village just before twilight and Aysha remembered that 'white medicine spread over the village.' The little boy was still in Aysha's arms. She covered his face with a wet towel but forgot to do the same for herself. Most people jumped into the river to protect themselves but Aysha didn't have time. The planes came to the valley and started shooting at people.

Aysha jumped from one side to the other as bombs exploded around her. Although she felt ill she did not realise that she was affected by the gas. She was too 'exhausted, hungry, and in pain' to realise what was going on. At dawn they arrived at Metin Mountain and a jash leader told them that it was no use. It was best to surrender. They turned themselves in and the men were separated from the women on site. All their belongings were taken from them and they were trucked away to Sersing and then to the Duhok fort. On the third day Aysha's face started showing the signs. She had blisters around her mouth, her eyes were oozing with a yellow liquid and she was blinded. That was the last time she saw her four brothers who disappeared without a trace.

Many women died and were injured during the different stages of Anfal. The majority of the wounded from the first Anfal ended up in Iran while others went into hiding or were captured by the government. Hundreds were affected by gas during the final Anfal in the Badinan region. Some of the wounded ended up in Turkey. This group, who lived in Turkish refugee camps, were visited by journalists and International NGOs to determine what had happened to them. These victims were the source of a small outcry at the end of Anfal when various organisations

---

15   Aysha, February 2006.

and countries asked the UN to investigate the use of chemical attacks by Iraq (see Chapter 1). Unfortunately this investigation was never carried out.

Health Problems

Women who survived the 1987 and 1988 gas attacks are now suffering from the delayed effects of the gas. The immediate health effects of mustard gas on the eyes, skin, and lungs are well documented (Romano and King 2001, Hashemian et al. 2006, Balali-Mood and Mousavi 2008, Ghazanfari et al. 2009). The long term effects include chronic lung disease and bronchitis, permanent impaired vision, and in some cases cancer (Balali-Mood and Mousavi 2008, Gosden and Amitay 2002). Nerve agents can cause muscle twitching, convulsions, contraction of pupils, excessive salivation, and involuntary urination. People who were exposed to nerve agents were regularly described as 'going mad' because they broke out in convulsions and some of them laughed hysterically. As yet, however, the long term effects of nerve agents on humans are unknown. Those who are exposed either die on the spot of appear to recover fully without long-term health effects (Asukai et al. 2002). In the following sections I will address women's ill health, their beliefs about their health problems, stigmatisation in the community, and psychological problems.

Long Term Illness

The majority of the women, even those who had minor injuries at the time, suffer from health problems. Some of this is clearly visible when you meet these women. They have aged prematurely. Some of them have sustained dark pigmentation on their faces, necks, and hands where they had been blistered by the gas. Some are continuously coughing, blowing their nose, and have difficulty breathing. They complain of exhaustion, fatigue, depression and helplessness. Fifty-three year old Sabiha had just got out of hospital two days before I interviewed her. She had been hospitalised for two weeks because of 'another chest infection.' Her voice was husky when she spoke. She had been informed that her throat and nose membranes had lost their natural moisture which made her prone to infection and coughing. She also reported suffering from various skin rashes and allergies. Her face showed signs of premature ageing and she felt depressed about her health.

At the end of April 2009 when I visited a monument-photo gallery of the victims of Balisan valley gassing in Shekh Wasan. I came across a picture of Sabiha. She had passed away on 19 August 2008, two years and four months after I had interviewed her, and twenty one years and four months after being injured by gas. When I asked Omar Babakir, who has put the photo gallery together, what Sabiha died of, he told me that 'she had no oxygen left in her lungs.' This layman's expression may refer to some scientific truth. 'Mustard gas causes extensive damage to the cells linking the entire respiratory tree,' wrote Professor Dlawer

Ala'Aldeen[16] in an email 'Depending on the dose and length of exposure, the damage can be severe, deep and destructive. After recovery, the damaged tissue will be replaced by non-functioning scars. This means, breathing and oxygen supply would be severely hampered, particularly during exercise or even minimal physical activity. These can eventually lead to failure of the respiratory and/or cardiovascular systems. Of course, DNA damage in lung or other body cells can lead to various types of cancer, which can also be a cause of death long after exposure.'

This is confirmed by Balali-Mood and Mousavi (2008) who point out that mustard gas can cause 'a wide range of chronic health effects including chronic bronchitis, bronchiectasis, frequent bronchopneumonia, and pulmonary fibrosis, all of which tend to deteriorate in time.' Sabiha may have died of lung failure or lung cancer. All the time she was given antibiotics and pain killers which did nothing to help her health problems (see section on available aid, below).

Mehabad[17] who was injured during the same attack lost sight in her right eye because her skin and eyes 'were burnt by the chemicals.' She also complains of memory and concentration problems, chronic chest infections and general ill health. She too is repeatedly hospitalised.

In a study about the survivors of the sarin terrorist attack on the Tokyo subway in 1995 Asukai et al. (2002: 157) found that exposure to sarin can cause chronic effects on the central and peripheral nervous system. Mainly sarin exposure can bring about 'cognitive deficits such as memory, abstraction, sustained attention and speed of information processing on the central nervous system, and decreased vibrotactile sensation and nerve conduction velocity on the nervous system.' These in themselves can be distressing symptoms that further depress the person and may cause other health problems.

Beliefs about the Long Term Effects of the Gas

Women survivors attribute all their health problems to the gas exposure. They believe that miscarriages during and after the exposure, the slow growth of their young children, and their experience of temporary sterility are all caused by the gassing. Uzra was seven months pregnant during the April 1987 gas attack on Shekh Wasan village. Five weeks later she gave birth to a stillborn. She recalled that 'the baby died after the gas attack, I felt it.' The baby had stopped moving after the gas attack and she was bleeding until she gave birth to him. Uzra kept on bleeding for a year after giving birth. She was worried that there would be no blood left in her and that she would 'dry up.' It is unclear whether this miscarriage was caused by the gas attack itself or the shock and terror that she felt as a result of the attacks.

---

16   Dlawer Ala'Aldeen, professor of clinical microbiology.
17   Mehabad, April 2006.

During the interview Uzra kept coughed violently, threatening to throw up. The villagers sent me to her as the example of a woman who is too ill to lead a normal life. Her chest is the worst, she says, 'I cough blood. I bleed through my nose and mouth. My throat bleeds. I have no solution, no treatment.' She is too ill to work in the farm and feels hopeless about her condition: 'I was really ill then and I am still ill, I have never got better. There is no treatment for me.' At the time of the attack she sustained many blisters and was blinded for about two weeks. She was unconscious for most of the first week and was very vague about where she had been taken and who had helped her. She now has dark pigmentation in the places where she had blisters and she looks a lot older that she is.

Fatma[18] had a baby three days before Ware was gassed. She was unable to leave the house or head to the spring like the other villagers. Ironically, this saved her life because the spring had been gassed and all those who washed themselves in the water died instantly (see Chapter 2). She stayed at home with her husband and seven children. He had built a large fire at the entrance to their house (which they had been told is a good way to fight a chemical attack). The whole family was injured by gas though this was not apparent at first. The baby who was only three days old is now unable to talk properly (18 years old at the time of the interview). He is in his mother's words 'dumb and dizzy.' Fatma believes that her son's slurred speech and slow growth is another consequence of the gas attack.

Khucha, who was exposed to gas attacks twice (see above), was unable to have children for ten years. She had three operations in which her womb 'was cleaned up' because 'it was full of dark stuff.' She believes it was the gassing that prevented her from becoming a mother even though she had been married two years before Anfal and had not been able to have children. Since the long term effects of the gas are unknown, survivors attribute all their health problems to the exposure. More research is necessary to determine whether some of the self reported health problems are caused by the gas, which types of gas cause what types of ailments, and how these could be addressed medically.

There is evidence that even though some physical symptoms may persist years after exposure, some of these may not have an objective ground. Asukai et al. (2002: 160) found that five years after the sarin attack in Tokyo physical complaints related to the eyes and general weakness 'tended to linger as subjective complaints ... however clinical examinations found no pertinent objective signs of these lingering subjective complaints.' The authors go on to explain that the long term health effects of nerve agents such as sarin have not been determined. This causes anxiety in survivors and it intensifies their subjective complaints.

Stigmatisation

Gas survivors are generally stigmatised in the community. Todeschini (2001: 108) speaks about the women survivors of the atomic bomb who were considered

---

18   Fatma, April 2006.

polluted and possibly passing on their contamination to their offspring and other members of the community. Hence Todeschini (2001: 109) speaks of the Japanese society's 'intense fear of contamination' of atomic bomb survivors. This fear was more apparent in the case of women who bore signs of having been victims of the atomic bomb. Similarly, Romano et al (2007: 653) speak of how survivors of chemical weapons may feel 'feared, unsupported, rejected, ostracised and even ashamed.' This is confirmed by Asukai et al. (2002: 157) who found that individual survivors of the sarin attack on Tokyo are 'stigmatised as a contaminated person.' This contamination, the authors go on to explain is not merely physical but also spiritual and this causes major distress amongst the survivors.

Khucha suffers from repeated chest infections, skin rashes and vision disturbances. Throughout the interview Khucha's nose was streaming and she spoke about her 'shame' for having to continuously blow her nose in public. She feels stigmatised in the community as none of her neighbours help her because they think they will catch 'the disease.' She complained that she had been in bed for ten days and not one person from the village had gone to check on her or 'give me a glass of water', an expectation which still holds in the villages where people look out for each other at difficult times. 'We were gassed 15 years ago' Khucha complained, 'but even now people don't dare to come close to us. When I fall ill they tell each other: Don't go, she smells of the chemicals. They think they would catch it from me.' She also suffers from muscle pain which prevents her from working. She and her husband end up in bed for long periods, unable to look after themselves or their children. She is fed up with taking numerous pills and injections most of which seem to be ordinary pain killers and antibiotics.

Psychological Problems

Gosden (1998b) speaks of the difficulty to quantify and diagnose neurological and psychiatric problems. These problems make people feel extremely desperate. They are also difficult to treat because antidepressant drugs may have severe side effects on survivors of nerve gas. Layla who survived the 22 May 1988 gas attack on Ware (Anfal 5, 6, 7) was acting deranged when she regained consciousness after the attack. She could not see and all her body was aching. Her skin was blistered. She could not breathe, and her eyes and nose were streaming. She kept shouting and crying, confused about what was wrong with her. She was told by relatives that she had been injured by the gas attack but she could not understand what they were talking about, 'What gas attack?' she kept asking them even though she had witnessed the death of some 20 people in the village during the attack. Other people in the village reported that for a few years after the attack Layla was considered mad, she was odd, depressed, and confused.

Some of the psychological problems may be the result of physiological changes brought about by the attack (more research is required in this field) whereas others may be a normal reaction to acute trauma. The trauma of living through the attacks, wtinessing grotesque death, and losing members of the family are enormous sources of suffering (see Chapter 6). These psychological symptoms may in turn lead to more physical symptoms and hence the circle of hopelessness and illness continues. In addition the trauma associated with exposure may be easily triggered with minor eye problems or general daily exhaustion. It is, therefore, difficult to distinguish between physical effects of the gas from the psychological outcomes related to it.

Romano et al. (2007: 638) confirm that psychological complaints continue in survivors, irrespective of acute physical effects. This, the authors argue, suggests that other factors such as the stress associated with a chemical attack may have prolonged psychological effects. Stokes and Banderet (1997: 405) point out that high levels of stress, associated with the unfamiliarity, unpredictability, ambiguity of the chemical and biological weapons' long term consequences, and helplessness of being exposed to gas attacks can cause changes in the 'physiologic, neuroendocrine, and immune systems of the body.' Thus can result in real illness. The authors then go on to state that it is important that survivors understand how stress can actually cause the worsening of their symptoms through a physiological process which causes real physical symptoms.

There is a sense of insecurity and fear as women feel that they may fall ill at any time, they may develop cancer and other rare diseases. This fear is comparable to that of atom bomb survivors who felt that their bodies 'could turn against them at any time' (Todeschini 2001: 106). Since gas survivors don't understand what is caused by the gas they count all their problems as side-effects of the exposure. Todeschini (2001: 121) speaks about the psychological suffering of women survivors of the atom bomb for whom: 'the very uncertainty about his or her future, including that of offspring, is a source of enormous suffering, and thus devastatingly real; but such suffering has no place in the biomedical model.' This means that there is pressure on atom bomb survivors to 'overemphasise their bodily complaints in order to receive treatment at all.' The psychological problems in themselves were not considered 'real illness' (Todeschini 2001: 123).

Social Consequences

The gas exposure may have affected some women's chances of getting married and having children. When talking about women who were children or adolescent at the time of Hiroshima and Nagasaki Todeschini (2001: 103) argues that these woman are 'seen as tainted by radiation.' Contrary to gender expectations of nurturance and child bearing the atomic bomb survivors are seen as 'undesirable others' in the Japanese society. Similarly, some girl survivors of the gas attacks may have been unable to start a family because they were ill and unable to nurture others. The worry associated with not being able to give birth, or giving birth to 'deformed' children has made some of the Anfal survivors undesirable as marriage partners.

Generally speaking the rest of the population seems to be fed up with these women's continuous complaints about their health. In the villages I visited, people seemed to want to suppress these stories of pain, desperation and hopelessness. On two occasions when I interviewed gas survivors people who were sitting in the same room sighed and walked out. On one occasion my gate keeper openly cut a woman short when she talked about suffering and not having access to any support. He told her: 'Thank you very much, you have made your point.' This is in line with the experiences of the women survivors of the atom bomb who were 'accused by exasperated doctors of being neurotically preoccupied by their illness and fears' (Todeschini 2001: 122). This, however, blames the victims for what they are experiencing. It distracts attention from the historical and political issues and from the failure of the Kurdish government and other international bodies to respond to the needs of these women.

Available Aid

In early 1998, Gosden along with the Washington Kurdish Institute established a Post-Graduate medical program for research about and treatment of chemical weapon survivors in Iraqi Kurdistan. This research project linked the Medical Colleges in the three universities of Duhok, Erbil and Suleimanya. One of the aims of the project was to raise support for the survivors by investigating the long term effects of WMD and developing new treatment methods. It was also hoped that by forming a national and international collaboration between academic hospital-based centres, community based programs, NGOs that would provide humanitarian relief the exposed population would be better supported. A year later in an article published in Washington Post Gosden pointed out that 'these plans have floundered on the rocks of political expediency, bureaucratic malaise and institutional rivalry. I realise now that although impassioned, I have been ineffective and have failed to provide effective help and support for the people of Halabja and other WMD survivors.' She goes on to state that this is because 'no major assistance for a medical/humanitarian program has yet been made possible. Compassion fatigue, cynicism and red tape abound, despite sympathetic supporters in government and among NGOs keen to help.' (Washington Post 10 March 1999). This, combined with her own lack of scientific credentials on the use of gas in warfare, meant that no reliable research has been forthcoming.

The gas survivors in Iraqi Kurdistan are angry that nothing is being done to help them. On a number of occasions women who were desperately ill came to see me and asked me for help. I had to repeatedly inform people that I was there as a researcher and I was unable to help them in any direct way. A woman told me she does not want to talk, she just needs help with her health problems. Another refused to speak with me, stating: 'What is the point? No one will help.' Many people feel used by researchers, journalists and even government workers who seem to be collecting data and making documentaries and reports without any help materialising (see Chapter 6).

At the time of the interviews none of my informants had received help from the government and they had all paid for their own hospitalisation and drugs. Many needed continuous care but were too far from the cities where the main hospitals are. Sometimes, they reported, they just stayed in bed in their distant village and got depressed. Khucha paid for her own hospitalisation which was very expensive. Each time she needed to see the doctor she had to pay for transport to Suleimanya, the nearest city, and then pay for her tests and medications. She complained about lack of financial support from the Kurdistan Regional Government and told me how when the peshmarga were fighting the government she baked bread for them, washed their clothes, fed and sheltered them. Now no one cares about the fact that she is 'dying slowly.'

Sabiha was also disillusioned with the Kurdish revolutionaries whom she had served throughout the 1980s. She felt that she had helped and supported the peshmarga when she could but when it was their turn to help and support her, it never materialised. She was fifty three years old at the time of the interview and had been gassed 19 years before (see above). She complained that the Kurdish government and parties had not done anything to help the gas survivors: 'Not even giving us an injection, all these years.' 'Nothing,' she stressed, 'No house, no medication, nothing.' Despite her bad health and withstanding the traumas of being gassed, losing her husband and son, she worked as a cleaner in the primary school in Balisan: 'If I was not desperate (for money) I would not work at this age and with this health. I have been ill for two weeks, I was in hospital. I keep falling ill.' Unfortunately, Sabiha died of illness two years after I spoke with her. Her expectations and hope for support never materialised and she died slowly, in agony, and disappointed with the ideals she once believed in.

Shler[19] from Balisan complained that despite being interviewed by various people every day, no help has arrived for her family of seven (she was injured with her husband and five children in the 1987 gassing). 'We are always ill,' she said, 'always in hospital. We have had no support, not even a cream or an injection … my husband needs an eye operation which cannot be done here … we have chest infections, vision problems, skin rashes.' Even now the Kurdistan Regional Government only supports women survivors who have lost their husbands, brothers or fathers to Anfal. Women whose husbands have survived the gas attacks and are continuously ill receive no support. They are also unable to work because of their bad health and are amongst the poorest and most isolated members of society vulnerable to depression and other psychological disorders.

Conclusion

Chemical weapons can have a devastating physiological and psychological effect on those exposed. Some of the short-and-long term physical consequences of

---

19   Shler, April 2006.

these weapons are well documented while there is still uncertainty about other aspects of how mustard gas and nerve agents affect the physiology of the body. The mutagenic effect and the typology of some of the deformities that survivors experience in Iraqi Kurdistan require further investigation. On a psychological level the intense helplessness caused, and the unpredictability of these weapons makes them powerful weapons of terror. This in itself can contribute to the continuation of physical symptoms, further ill health, and uncertainty.

The gas survivors in Iraqi Kurdistan have received minimal support from their own community, government, and the international community. The stigma experienced as a result of being 'contaminated' means that survivors are deprived of community and social support which is a major adaptive response to trauma. They have also fallen into a blind spot as far as the Kurdish government's aid programmes are concerned. Their specific needs for medicalisation, counselling, and reassurance have not been met. The international community has not yet acknowledged its responsibility towards the survivors who are victims of not only a brutal dictatorship but also of Western companies that sold chemical precursors to Baghdad throughout the 1980s, often with the knowledge of Western governments in whose countries these companies were based or across whose territory the materials were shipped. An acknowledgement of responsibility and attempts at redress are crucial as the survivors of these attacks continue to suffer and some of them die a gradual and painful death.

There is a need for educating the community and the survivors themselves about how the gas attacks have affected their health and mental health. There is little mental health care in Kurdistan and few psychiatric units if any. Women who suffer from depression and other psychiatric problems have no means of accessing help. Contrary to women survivors of the atom bomb who receive medical relief the gas survivors in Iraqi Kurdistan receive no help from any government or organisation. They over-emphasise their physical problems in order to get some form of support but so far no specialised health care is available for them.

Chapter 5

Rebuilding Life after Anfal:
Employment, Poverty and Exploitation

We worked, laboured, ran... earned money and spent it on food.
(Behe, November 2005)

On 6 September 1988, a General Amnesty was announced to include everyone except Talabani (in effect it also excluded the Assyrian, Chaldean, and Yezidi communities). The women detainees were released from the camps. Some of the refugees returned from Iran and Turkey (the majority stayed put until 1991 when the UN No Fly Zone was put in place) and those who had gone into hiding during the campaign came out and surrendered to the government. Only then the women realised that their husbands, brothers, sons and, in some cases, entire families had disappeared from the community.

Until 1991, the whole region was under Iraqi control. The women had no Iraqi citizenship, because of not taking part in the October 1987 National Census. As a result they had no entitlement to food rations. Their children were not accepted in schools and they had no income. They were not allowed to return to their levelled villages but were forcibly located in the various housing complexes at the edge of cities. These were built for the deported rural communities throughout the 1970s and 1980s. Some of the villagers, in particular the inhabitants of the Badinan region, were dumped in an empty desert that had no housing and no community.

In the aftermath of the amnesty everyone was being watched by nearby military bases and they had to report regularly to prove that they were still where they were supposed to be. After about two years, and especially after Iraq invaded Kuwait on 2 August 1990, the strict surveillance on the villagers alleviated. People were allowed to move about more freely and to move to other areas where they had family and support.

Rebuilding life after Anfal was a major challenge for women. The majority experienced a radical change of status when they became the sole breadwinner in their families. Most of the women were uneducated and had no transferable job skills. They ended up doing the lowest paid jobs that no one else was willing to do. They worked as porters, labourers, servants, bakers, builders and factory workers. The hard physical labour took years from their lives. Over the years some women have developed health problems which prevented them from continuing to work.

In 1991, everything changed once again for the survivors. In the aftermath of the first Gulf War, and encouraged by America, the Shiites in the south and Kurds in the north launched the Popular Uprising against the Iraqi state. Within weeks the uprising was crushed by the Iraqi army and this led to the mass exodus of

Kurdish civilians into Iran and Turkey. Two million people fled to the borders at the end of March and beginning of April that year.

The images of desperate refugees battling with rain, hailstone, wind and hunger were broadcast on international channels. The civilians stayed in tents for a few months until the No Fly Zone was set up to protect the Kurds from Iraqi attacks. Anfal survivors, who had fled to Iran and Turkey along with the rest of the community, returned to find that the minimal belongings they had put together over the last two and a half years – blankets, clothes, pots and pans and bits of furniture, had all been looted. Once again they had to start from scratch. Their return also coincided with a double sanction on the region, the first imposed by the international community on the whole of Iraq and the second imposed by Iraq on the Kurdistan region.

In 1992, after several minor armed skirmishes the Iraqi army withdrew from the Kurdish region (except from the oil rich regions of Kirkuk, Khanaqeen and Singar). In retaliation the government withheld salary from the civil servants in Kurdistan and imposed a sanction. At this stage the first Kurdish elections were held in which Anfal survivors took part with great joy. It was the first time that people could vote freely for a group which, they believed, represented their interest. The establishment of a Kurdish government and the end of Saddam Hussein's rule was celebrated in the whole of southern Kurdistan. Everyone was hopeful about this new era of freedom. It was hoped that a new society would be built which would be free of oppression and violence.

Economic hardship affected the whole of the population in the following years. Everyone struggled to survive and there was no major difference between rich and poor. This equity in hardship was maybe the reason why the society was more united and there was no resentment at being deprived and marginalised. With the rapid involvement of International NGOs and with help from Kurdistan Regional Government (KRG) thousands of villages were gradually rebuilt in the early to mid-1990s. Many Anfal surviving families initially returned to their villages but left once again a few years later as a result of urbanisation, drought and civil war (see section below).

The Kurdish civil war broke out in 1994 and finally ended in 1998. These were the most desperate years when the KRG's limited resources were wasted on the war. More young men got killed in the process. Some areas, which were war zones between the various sides, became uninhabitable. In August 1996 and as a result of a major battle between the two parties, the KRG was split into two administrations. The PUK ruled in Suleimany and Garmian and the KDP ruled over Erbil and Duhok. Each of these parties was harassing and evicting members and supporters of the other party in its own region. Generally, there was a lot of nepotism and those who supported neither party struggled to find employment and to have access to resources.

The UN's Oil for Food programme in 1997 gave Iraqi civilians basic food stuff in return for Iraqi oil. The Kurdistan region was assigned 13 per cent of this programme's import even though the Kurds comprise 20 per cent of the Iraqi population. Despite this the programme helped alleviate the general poverty and provided basic living security to civilians. The major change, however, came in the aftermath of the 2003 invasion of Iraq. With the lifting of sanctions on Iraq and the establishment of a new Iraqi government, in which the Kurds were one of the key players, it was as if the flood gates were opened. Suddenly a lot of wealth poured into the region.

In the aftermath of 2003 the Kurdistan region looked like a building site. Money was dedicated to regeneration and infrastructure. The possibility of importing goods from neighbouring countries developed rapidly. The various business deals and contracts led to the creation of a wealthy and advantaged class, most of whom were associated with one of the major parties, in contrast with a poor and marginalised group. The Anfal survivors remained poor and the Anfal affected regions remote and forsaken. This is when resentment grew against the Kurdish government and survivors started demanding their rights in an angry fashion. In this chapter I will present some of the important issues related to survival and rebuilding life, as recounted by the women themselves.

Accommodation

Since the Anfal campaign, accommodation has been one of the major problems for women. The village houses had been modest but comfortable. They were made of mud brick designed to be cool in the summer and warm in winter. The rooms were decorated with red Kurdish carpets and rugs, and people sat on the floor, leaning on round cushions. Each family had its own private garden and orchard. The move from a house with a garden to a room which may not be in a good condition is likely to erode the individual's happiness and wellbeing. Anfal destroyed the notion of home, privacy and being a host. In this region having a house is closely linked with being a host. The gracious social traditions of having guests, of hospitality and connectedness were destroyed by the campaign. It also destroyed the sense of safety, routine and sense of belonging that comes with having a home and being grounded in a community (Bassuk et al. 2003: 42).

At the beginning and for a long period most women ended up living with their extended families and relatives before they were able to find a place for themselves. Most of the time, the accommodation consisted of one or two rooms where a few families, with large numbers of children, shared the space. Everyone slept alongside each other. There was no privacy and no space to rest. Many women ended up moving a number of times, living with various relatives and strangers. Having come from beautiful villages surrounded by hills, mountains and cliffs the women found their new, overcrowded, flat and dusty neighbourhoods difficult.

Even when they were able to move out of relatives' homes and become more independent many of the women ended up living in huts and sub-standard housing that had no water supply or electricity. They had to go to the water source several times a day and bring home heavy water containers on their heads and shoulders.

Sometimes fights broke out in the water line. It was particularly difficult for young girls whose mothers worked. They had to queue for water for a long time and sometimes fight to get some home. The following women ended up living in Smood and Sarqalla (in Garmian), Shorsh (near Chamchamal, Suleimanya), and Bahirka (outside Erbil).

Khursha[1] reported living in a '*kalawa*' – a ruin or a piece of land where waste is dumped – while she lived in Smood housing complex. She reported having 'no clothes, no furniture, nothing.' The place did not have a water tap and she had to walk a long way to bring water from the source. She believes that if it was not for the support provided by family, relatives and complete strangers she and her two children would have starved to death. Similarly Rezan[2] reported having nothing to live on in Smood. The family 'were extremely poor and needy.' They only had what people gave them 'as charity.' According to Rezan, Smood was 'a ruin, an awful ruin.' The roads were not paved which meant that there was a lot of dust in the summer and mud in the winter. Hundreds of desperate families from Garmian ended up in Smood. Those who were already living there, as a result of earlier deportations in 1987, ended up accommodating many relatives and acquaintances. Many survivors, however, ended up sleeping rough, under tents and in ruins.

Both Khursha and Rezan ended up leaving Smood and settling down in Sarqalla. They decided to move to Sarqalla because 'it had water and there was work.'[3] Also because, Rezan pointed out, the people were gracious and they helped the survivors. They made sure the women were included in society and were given work: 'Whenever they had work they asked the Anfal women to do it, they wanted to help us.' Khursha feels grateful to the people of Sarqalla, 'from the young to the elderly,' because they have 'respected us and been kind to us. They have helped us as much as they could.' She points out that it is not just about giving money to those in need, but more than that, it has been about respect for her: 'Respect is more important than everything else.' After seventeen years of renting and suffering, both women were given a house each by the Kurdistan government in 2005. This has made all the difference to their living standard and their wellbeing.

Shadan[4] initially lived in Chamchamal but because of a lack of work she went to the Arab district of Hawija, near Kirkuk, where she worked as an agricultural labourer. After seven years of hard work the family returned to Kirkuk, Shadan's home city, but eight months later they were deported again as part of the Arabisation of the oil rich areas. In 2000 she went to Barda Qaraman and stayed there for one year until the Kurdish government set up a camp for them in Takya. This is where she stayed 'until the liberation of Iraq in 2003.' Then, in an attempt to normalise the Arabised regions the KRG brought some Kurdish families back to Kirkuk. Shadan lived in a tent: 'They dumped us by the stadium.' The family

---

1   Khursha, November 2005.
2   Rezan, November 2005.
3   Rezan, November 2005.
4   Shadan, March 2006.

'built half walls and covered the top with canvas' and this is where Shadan ended up living for two and a half years. In 2006 there was a flood, Shadan's furniture was drowned and the walls collapsed. When I interviewed her she was living in a neighbour's house and 18 years after Anfal she was still without a home: 'We need to find accommodation.'

Shamsa[5] reported staying with various people until 1991. Initially she was hiding with a family in Kirkuk (see Chapter 3, section on IDPs). After about six weeks she returned to Shorsh housing complex near Chamchamal where she stayed with her parents in law who lived in the open. She was then given some material by people and she built a house. She built half walls and made a roof from 'an old door which had been thrown away.' In Chamchamal they had 'nothing, no money and no belongings.' The house she built had no toilet and no water supply. All around them there was just 'dust and dirt.' Many people, especially the elderly who had no young members of the family to look after them, ended up sleeping in the open 'and there were snakes and hungry dogs around you.' In a similar fashion to Shadan she returned to Kirkuk only to be deported. She then ended up working in Hawija for a number of years and eventually she remarried and moved to Badawa, her husband's village.

The villagers from Badinan region were dumped in Bahirka, outside Erbil. It was 'a flat desert, ugly, all thorns,' recounted Adiba from Derkari Ajam.[6] She believed that they were brought there to die: 'They dumped us there and gave us nothing ... There was no food, no water, no shelter.' Fairooz,[7] a relative of Adiba who ended up in the same place, pointed out that when they arrived at Bahirka there was 'no one, no settlement, no tents, nothing.' Everyone ended up living in tents provided by relatives who had come to visit from Badidnan and by the people of Erbil. In the winter the tents were battered by the rain. It was cold and all around them it was mud. There were no toilets in Bahirka. After a few weeks some people, according to Nalia,[8] set four wooden pillars on the ground, covered them with the burlap sacks and blankets provided by the people or Erbil and they made rest rooms. That year in the desert of Bahirka, which was known to be dry and lifeless, a water spring erupted, in a biblical manner, which helped the inhabitants to survive.[9]

Sometimes the inhabitants of Bahirka managed to smuggle petrol in but they risked being beaten up at the checkpoint. Most of the time, however, the people of Erbil brought them the necessary things to survive.

Nalia, whose husband was one of the 32 men shot in Koreme, lived alone in a tent with her two little sons. Some nights she was too scared to sleep. Once she dimmed her lantern and slept 'under the cradle' hoping that if anyone came

---

5   Shamsa, March 2006.
6   Adiba, February 2006.
7   Fairooz, February 2006.
8   Nalia, June 2010.
9   This was confirmed by villagers from Koreme, Chelke and Derkari Ajam.

into the tent they would not notice her. She was scared of being attacked because so many strangers lived in the camp and she was a lone woman. After a while a relative let his grown up daughters stay with her. Nalia's tent was covered with a nylon sheet to shelter her from the rain. Stones were secured to the edges of the sheet to keep it down. One cold and windy night the stones were constantly being lifted up and they banged on the canvas, making sounds which terrified her. She thought someone was trying to get into her tent.

Accompanied by one of the young women who stayed with her Nalia went around the tent but found nothing. After going back inside the banging started again. She then took out a knife and went around the tent once more. The three women then decided to go to a neighbour's tent because they were too scared to sleep. The neighbour, who was also a widow let them in and comforted them. Soon the Imam called for the dawn prayer and Nalia felt relieved because the sound of the Imam promised sunlight and a new day buzzing with people and life. A few months later, helped by strangers and relatives, Nalia was able to build 'a small hut' and she stayed there until 1990 when she returned to Duhok and moved in with her parents.

In the spring of 1992, a team from Physicians for Human Rights along with Human Rights Watch uncovered the mass graves in Koreme and unearthed the bones of her husband. Nalia then decided to go back to the village: 'It gave me solace to be close to his grave.' In Koreme she lived in a tent with her two children. Few people had returned to the village as the conditions were really harsh. One windy night Nalia was too scared to sleep. The wolves were howling nearby and she felt vulnerable and lonely. She decided to return to Duhok few weeks later. After about two years an NGO built some houses for the villagers in Koreme and Nalia decided to go back once again.

The new houses had many problems. The roofs were particularly badly designed: 'Snow and rain were pouring into the house in every direction.' She did not know how to fix the roof and there was no one to help her. Traditionally fixing roofs is a man's job. Having grown up in a city and being without a man Nalia was unable to deal with these problems. Finally, after one year, she moved back to Duhok for the second time and once again lived with her parents. In 2001 Nalia was eventually given a house by the Kurdish government and she now lives there proudly with her two sons.

Survival was particularly difficult for young widows with little children and older women who had lost their husbands and sons. Nalia's aunt lived in a tent in Bahirka with her four daughters. One harsh winter night when the abundant rain and snow made the tent heavy she had to hold on to the pillar in the middle of the tent to prevent it from collapsing. The palms of her hands were rubbed raw by the next morning. She had not slept all night. Later her brother-in-law, Nalia's father, brought the old woman a sheet of nylon to put around her tent.

After the Popular uprising in 1991 when the Iraqi intelligence, security and police retreated from the Kurdistan region many of the buildings of oppression were deserted. Soon Anfal survivors and other internally displaced people occupied these buildings. In a visit back to the Safe Haven in 1995 I went to the Security building, known as Amna Sureka – the red security, in my hometown of Suleimanya. The place had been used for torture and interrogation during Iraqi rule and many of the prisoners had carved their names on the walls.[10] I found dozens of large families crammed into small rooms. They lived side by side with the engraved names and drawings of prisoners who had left massages for their families before they were killed. The place was overcrowded and desperate.

In the city of Suleimanya the internally displaced were also living in the ruins of an old hotel called Haseeb Salih. Those who lived there were stigmatised by the rest of the community. They were called 'kotrekani Haseeb Salih' – the pigeons of Haseeb Salih. They were perceived as uneducated, backward, and tribal people who did the lowest paid jobs and whose children begged on the streets. Ironically, in the late 1990s the building was renovated and now it has become one of the most expensive hotels in the city. Visitors who stay don't realise the history of that place and the misery of the conditions of living when it was full of people struggling to survive.

The majority of the forts, military bases and prisons which were used as assembly centres during Anfal were also inhabited by internally displaced people, some of them Anfal survivors. The Nizarka fort in Duhok, Topzawa Popular Army Camp, and Duz Khurmatoo Youth Centre are still full of such people. The larger halls are divided into smaller rooms, usually separated by half walls on either side of which a different family lives. These inadequate settlements lack basic urban infrastructure. They have primitive sewage which places the inhabitants at risk of illness, especially in the summer when there are mosquitoes and flies. Many of them don't have a main water supply, electricity or telephone lines.

There is evidence that chronic bad conditions such as living in substandard housing, being poor, having taxing responsibilities and experiencing discrimination lead to mental and physical health problems which in turn lead to more losses and more economic and social insecurity (Belle and Doucet 2003: 102, Myers et al. 2005: vii). They also lead to feeling of isolation, damaging 'the sense of connection between individual and community' (Herman 1992: 55). This is especially true when survivors live in a wealthy society and there is a large gap between the rich and poor (Bassuk et al. 2003: 33). The sense of injustice experienced by survivors as a result of the majority's lack of interest in helping them damages society cohesion. It also creates a deprived underclass that has no voice and no opportunity to improve their situation.

Employment

After getting over the initial shock of finding themselves displaced, without men, and entirely responsible for their children and elderly dependants, the

---

10   Now it has become a museum were some of records of oppression are being kept.

women survivors set out to work. Despite their illiteracy, the limited employment opportunities for women, and the difficult social and political situation, they proved to be extremely resourceful in finding work and making ends meet. In the spring they went to the mountains and hills all day, gathering vegetables and herbs for their own consumption and for selling. They produced brooms from straw gathered from the fields and sold them on the streets. They knit and sold lifka, a hoarse traditional flannel made from sheep wool. They took on sewing, baked bread, became cleaners in institutions and worked as labourers in factories and on other people's land. They also collected the few wheat flowers that were left after the harvest and ground them to make flour.[11]

Coming out to work was out of line with the traditional system they were familiar with. In the tribal system women are provided for and they are expected to stay at home, to look after their children and to protect their husbands'/ brothers'/ fathers' honour. Women who came out to work faced social and tribal problems. They were stigmatised and targeted by gossip and social pressure. Mlodoch (2008) points out how these women were perceived as 'women at risk' because they had no male guardian and they were getting out of their homes to work. Some were even victimised by honour killing and domestic violence. Here I will talk about the different kinds of work women carried out and how it affected their health. I will also talk about their resort to illegal work and the general stigma that surrounded them in the community.

Being Resourceful

> There is no work left that we did not do – baking bread, weaving wool, labouring jobs, harvesting ... we did everything. (Peri, November 2005)

Shamsa reported going to the hills early in the morning and digging all day to collect 10 kilograms of kingir. This vegetable grows in the ground and its top layer is thorny leaves. You have to dig quite deep to get to the flesh of the vegetable. Collecting kingir all day put strain on Shamsa's shoulders. She spent hours putting her head down and digging for little money. She sold each kilogram for one dinar: 'We sold it on the road, people would buy it and take it to the cities.' This, however, was only possible in the spring. Other months of the year Shamsa 'knit tassels and decoration for scarves and shawls, and did some sewing.' Later, when she lived in Hawija she became an agricultural labourer, planting wheat, barley, cotton, and sesame. She worked all day for 5 dinars (approximately equivalent to 5 dollars at the time).

Lana,[12] who unlike the majority of the women had had some basic education, was 'really depressed' after Anfal but had to work to support herself. She started

---

11   Gulla Chini is something that many women referred to, it means collecting the left over flowers and crops.

12   Lana, March 2006.

work as a cleaner in a hospital in Kirkuk and, supported by her Assyrian manager who knew her story, she was put in charge of the sterilisation machine. After that she worked in the tile factory in Laylan for two years, doing 'a man's work,' using heavy machinery. She also worked in an ice-cream factory for a while. Then she ended up working in the poultry factory in Beji. Eventually, she joined her sister's family and went to Hawija where she worked as an agricultural labourer: 'I carried on like this till the liberation of Iraq [in 2003].' Then she started working for an Anfal women's organisation. Lana does not even benefit from the Anfal support salary because her younger daughter, now a young woman, receives that.

Behe[13] also reported planting and harvesting 'whatever was in season.' She had five children, two of whom were disabled. She left her children at home and worked 'from dawn till night.' She did whatever was available and she 'never sat down.' She was the angriest amongst the women I talked to in Sarqalla. She had moved house a number of times, lived in old and abandoned places which needed a lot of fixing: 'We died labouring for other people, making mud bricks, fixing holes in our homes.' She was angry that no one was helping her disabled son and daughter. They both have spine deformity and need a lot of physiotherapy in order to walk, something which she cannot afford. She was also angry that her labour, grief and pain meant nothing to the rest of the community; and because there is so much inequality in the community: 'the salary given to the Anfals (150,000 dinars at the time, equivalent to $100) is equivalent to what the wives of the senior politicians spend on make up.' Fortunately, in the last five years, this salary was doubled twice and this has been great help to the survivors.

Health Problems

The income women generated was barely enough to feed themselves and their children. Most of the women are completely exhausted. Their skin has been battered by the sun and rain. Their faces and hands are prematurely wrinkled. Many women reported having chronic aches and pains which prevents them from working now. Peri, who 'worked as far as [she] had strength', is now unwell. She has injured her back, neck and shoulders by the hard labour. She was in constant pain but she did not understand why. Recently, she visited a doctor who explained to her that it is 'over-exhaustion' because she has done 'lots of hard work for so long.' She thinks she is now 'paying for neglecting' her body for so long.

Working in environments without health and safety regulations and without wearing protection has also affected some women's health. Shadan worked as an agricultural labourer for seven years. She reported that her hands 'were rotting' by the end of this period. Despite being in her forties Shadan complained about arthritis, chronic chest infections, and deteriorated vision. She blames working on the plantations in Hawija for her infections and vision problems: 'I think it is because of the tobacco plantations ... the sulphur in the pesticides affected me.'

---

13 Behe, November 2005.

Asukai et al. (2002: 157) points out that Organophosphate agents are widely used as pesticide around the world and they may cause accidental poisoning. The pesticides can give off a gas which irritates the lungs and can cause infection and other health problems.

Shno and her sister Nasik[14] worked with their mother on the fields of Sharzoor. It was very cold in the mornings when they set out to work in the cotton fields. They had no money to buy socks or gloves, Shno pointed out, so they put 'plastic bags on [their] feet and hands' to keep them warm. Shno also suffered from malaria that year and she regularly had fever but she hid this from her employers. The women were paid according to the amount of harvest they made. They worked non-stop to pick another kilogram of cotton, okra, tomatoes. Nasik recalled that some days they did not have time to stop and have their lunch. This was usually a cucumber sandwich brought from home: 'Tomato was more expensive so we usually had cucumber instead.'

Shno and Nasik reported that once a beetroot field was flooded and the employers gave up on it. The women then took over digging the vegetables out, carrying them to the road and selling them. They worked long hours, bending over in the wet field until the tips of their fingers cracked and bled. Carrying heavy weights was particularly difficult. Their backs were bent under heavy sacks of food which they loaded into the trucks and tractors or carried all the way to the main road. Shno pointed out that the loads were so heavy that 'the distance from here to that door (5-6 meters) seemed like a year's distance.' Her sister, Nasik, suffered particularly from carrying heavy loads. For years she had stomach pain and she was told that she had hernia. She did not, however, have an operation because she could not afford it and also because she could not afford to stop working. The hard work continued even when they returned to their village, Kulajo. The sisters had to make mud bricks, carry them and rebuild their house. Nasik was in a lot of pain and she took to wearing a belt on the place where it was hurting 'to stop it from popping out'. She neglected herself and her need for an operation for years. A few years ago she finally had the operation. However she feels that she has damaged her capacity to work as hard as she used to.

Nalia grew up in Duhok. She moved to the village of Koreme when she married her cousin. Life was difficult for her because she had never done work outside the house. Her husband appreciated that she was a city girl so he helped her out. After Anfal, when she was a single mother with two children, Nalia started making Kurdish clothes for women and selling them in the market. She spent many nights sewing by the light of a lantern in the blackouts, using a manual and rusty sewing machine. Now, even though she is only in her early forties she cannot sew anymore. Her vision is very bad and she has injured her neck and shoulder as a result of looking down and straining her neck.

---

14   Shno and Nasik, May 2010.

Illegal Work

Some Anfal surviving women were forced by necessity to smuggle petrol and food through the various Iraqi checkpoints in the region. In view of the sanctions imposed by the Iraqi state the women risked being abused and sexually molested. A story that widely circulated in the community in the late 1990s was that of a woman who had smuggled petrol through a checkpoint near Kirkuk and a soldier poured petrol on her face and set it alight. Most of the time, however, the women were beaten and abused both as Kurds and as women. Sometimes they were sexually abused. This made them feel guilty and shameful about themselves and it also gave them a bad name in society. They were perceived as loose women who collaborated with Iraqi soldiers to make money.

A woman tried to explain this situation to me. This is how she described the smuggling work: 'We had no income, no salary, and no food rationing. The only thing was smuggling. You would buy a pot of oil from Chamchamal, you would go to the Kirkuk checkpoint. Sometimes the soldiers would beat you to death and take the oil. Other times they allowed you to take it through. You would sell it on the other side – the [Iraqi] controlled region. Some women had to have a relationship with the soldiers in order to carry on.' This woman then went on to say: 'My mum told me to do it but I could not.' Admitting to doing this work causes stigma. The woman now lives in a new area and wants to keep things 'clean' and simple.

Other women may have resorted to prostitution out of desperation, although it is extremely difficult to obtain any data about this issue. It is common knowledge within the community that such things have taken place and some women were killed by their relatives because of it or because of false rumours about alleged sexual misconduct. A man who used to be a shopkeeper told me that one afternoon an acquaintance brought a beautiful woman to his shop just before closing time. She was not an Anfal widow but a widow of another massacre. The man suggested that they have sex with her in the shop: 'You won't regret it,' he told the shop keeper, 'she hasn't had children, she is really tight.' The man had a fight with his acquaintance and gave the woman some money and told her to stop doing this. She was crying, 'What else can I do?' she asked him.[15]

Adalet Omar,[16] a woman activist, mentioned that women who were suspected of 'not being virgins' due to rape or prostitution were killed by their families. Rumours can be extremely dangerous. Disclosing such issues or talking about them can lead to someone's death. Women survivors who may know about cases of prostitution are not willing to talk about it. This silence is a form of protecting the woman involved as well as protecting the reputation of the whole group. Anfal surviving women are very sensitive towards bringing such issues up. They deny all involvement and some of them get angry about being asked. They feel that they

---

15 Conversation with Goran Babali, personal communications.
16 Adalet Omar, Erbil, March 2006.

are stigmatised enough in the community and talking of such issues just leads to more stigmatisation.

Social Stigma

Despite their hard work and their success at surviving and raising their children the women are devalued in society. Taban Abdulla,[17] who works in Kurdistan Woman's Union to support Anfal surviving women, spoke about a child survivor of Anfal who had been the first in his class but refused to go back to school. Taban visited the family in Garmian, trying to persuade and encourage the boy to return to school. She found that he had been insulted in school because of his mother's work. His classmates who were envious of his success talked about him in a degrading way and called him: 'The son of the broom seller.'

The kind of work women do becomes a part of their definition and identity. It is as if doing low paid work, being poor, and living in a hut reflects some failure in their personality, as if it is the person's own fault that they struggle to make ends meet (Kluegel and Smith: 1986). Hence instead of admiration and acknowledgement for the hard work they do, society looks down on them and their children. The children, like their mothers, are then labelled 'Anfal children' and this in itself is used to imply that something is wrong with them.

Adalat Omar pointed to a similar problem when she said: 'There are still people who look at the survivors in a degrading way, especially those [survivors] who have no man as the head of family. Most of the time when talking about women who don't have a patriarch, they are looked at as a bad example.' In the patriarchal society, to have no husband and no father is a marker of shame (see section on Exploitation below). Mlodoch (2009) speaks of the daughters of the Anfal survivors and how 'their fatherless youth is considered a spot on their virtue' and how many of the girls struggle 'to find a spouse.'

Being illiterate, living in poor districts with minimal living conditions, lack of male breadwinners and the low status jobs make these families socially deprived. Bassuk et al. (2003: 34) argue that social deprivation extends beyond poverty. They define it to mean lacking 'freedom of choice, opportunity, political voice, and dignity.' Similarly, Amartya Sen (1997: 87) summarised social deprivation as 'a deprivation of basic capabilities rather than merely as lowness of income' and 'lack of income can be a principal reason for a person's capability deprivation.' Sen quotes Adam Smith who refers to the importance of possessing basic, simple things such as a linen shirt or a pair of shoes that make it possible to 'appear in public' without shame (Smith, Wealth of nations, cited in Sen 1999: 74).

A lack of access to opportunity to better their lives socially, economically and politically marginalises these women in the community and they experience discrimination. Belle and Doucet (2003: 106) describe discrimination as 'a process in which the dominant group's privileges are maintained, at the expense of a

---

17   Taban Abdullah, Suleimanya, November 2007.

subordinate group or groups' and hence maintaining and exaggerating the existing social inequalities. This process can happen unconsciously when 'stereotypes distort our impressions of individuals.' Poverty can also challenge women's ability to abide by their own moral standards and expectations (Belle and Doucet 2003: 102) forcing them to resort to work that causes further stigmatisation and social problems.

Poverty and discrimination may also lead to mental and physical ill health (Blas et al. 2008: 1684, Myers et al. 2005: 5, WHO 2009: 15). This can happen either because of the hard work that exhausts people and wears down their resistance or because of the induced low self esteem, helplessness, and shame. Women who struggle to rebuild their lives and to look after their children as best they can experience work overload. They also struggle against stigmatisation and discrimination within their own community. Sometimes women internalise the community's perceptions of them and fail to see themselves as the strong survivors that they are (see Chapter 6).

Poverty and Motherhood

Socially, mothers are expected to be absolute nurturers that protect their children from harm, socialise them in the appropriate way, pass on culturally important norms, and make sure their children grow up to be good citizens with a good chance of success in the community. Swan (1998) provides an example which explains the social construction of motherhood in society. She states that there can be many reasons why a child catches a cold but the story that dominates for the mother is that of her negligence. This is constructed in the mother's life experience within her community where mothers are seen as nurturers that take absolute care of their children. Women are told that this is part of their nature as women and as mothers and those who fail to do what is expected of them and what they perceive to be their duty, feel guilty.

Excessive responsibilities and hard labour have meant that sometimes women survivors of Anfal have not been able to look after their children or be there for them as much as they would have wished. Being a bad mother is something many of the women talked and worried about. Some of them reported feeling guilty for leaving their young children behind when they went out to work, for beating their children, for not being able to support their children's schooling. Generally, they feel responsible for their children's lack of good employment prospects as adults. This inflated sense of responsibility and guilt is in turn reflected in their image of themselves as failures. They disapprove of themselves and sometimes fail to see how given the circumstances they have done very well and have been very good mothers.

Guilt

Shadan had young children and she used to take them to the field with her. She remembers their situation with great sadness. Her youngest son was a toddler. She left him on a soft patch of land somewhere and tended to her work. Occasionally she went back to check on him. Sometimes he would be asleep and she found that the flies were 'eating his face' or the ants were 'climbing his body.' He was covered in bites and he developed skin rashes and eczema. He was always crying. But she had no choice, 'I had to leave him and attend to work.' Similarly, Peri left her toddler in the shade, on a patch of grass in the field as she harvested. The child sometimes swallowed things, fell, cried but Peri could not stop working.

Shamsa left her 6 year old daughter at home alone when she went to work. The girl wore their room key around her neck and from 1991 to 1998 there was no one to look after her. Every day Shamsa made lunch for her daughter before she left the house. The child was too young, she 'didn't know anything.' One of the cheap meals Shamsa put together was to fry dates and add an egg on top. Once the daughter added turnip to her sandwich. Shamsa recalled this with sadness: 'She didn't know that things like that don't go together.' Every day the little girl would be sitting quietly by the door waiting for her mother to come home. Shamsa worried about her daughter and hoped nothing would happen in her absence but she had no choice because if she did not go to work 'we would have no food to eat.'

Behe started crying when she talked about her children: 'I would leave them from morning to evening and beat them if they wanted to eat too much.' The fact that she could not afford to feed them enough when they were young haunts Behe and many other women whose children cried for food, especially for luxuries such as meat and chicken. In some cases this led to children questioning their mother's love for them and defying her authority (Belle and Doucet 2003: 105).

Working Children

Some of the women who had larger families could not survive by working alone. Many children ended up working alongside their mothers. Ruqia's[18] family needed more support. She had seven children and very early on they 'all started working.' She then went on to say: 'My children were this small (gesturing close to the ground) when they started selling chewing gum [on the streets].' She also made them sweets at home which they sold. When I asked Ruqia whether her children went to school she stared at me with anger and surprise: 'I didn't send them to school, how could I? They were all hungry, what school?' Things got better for Ruqia as the children grew up and worked as labourers but for a number of years life was extremely difficult: 'Only we know what we suffered,' Ruqia said and then in a manner typical of survivors she immediately went back to the moment Anfal took place: 'No one could imagine, we were barefoot … it was pouring … we were

---

18   Ruqia, November 2005.

soaked ... some fell into the swollen river.' For many women the difficulties they face in their current lives are continuous reminders of the violence and rupture that caused their present day situation.

Peri's elder daughter did not do paid employment but she missed her schooling because she had to stay at home and look after her elderly grandparents. The elderly couple had returned from Nugra Salman prison and were in a bad way. Peri's daughter was only seven years old when she started cleaning the house, cooking and baking bread. This woman provided for her parents-in-law, who had no one else left, and looked after them until they died. She had no choice but to let her daughter look after the house as she went out and worked. In the scheme of things, the priority was immediate survival. Many women had no choice in this, working meant earning enough money not to die and this was the most important thing they could do for their children.

Children's Schooling

Due to their mothers' inability to support their schooling many Anfal surviving children ended up with minimal education. Asmar[19] has a daughter and two sons. Her youngest child suffers from severe Down's syndrome. She reported that her elder son was 'clever in school' but after passing his second year secondary school exams he gave up his schooling: 'I had no money to buy him a pair of plastic shoes to go to school.' Plastic shoes are the cheapest kind, made from plastic and with a plastic sole. Asmar's son then started work in a teahouse and was paid 5 dinars a day (equivalent to 5 dollars). The daughter also studied up to third year primary school but at the age of 9 she gave up her schooling and stayed at home to look after her disabled brother. She cooked and cleaned and she 'brought water in buckets on her shoulder.' Asmar feels guilty for not being able to support their schooling and blames herself that 'both of their futures were ruined.'

Similarly, Keejan's surviving five children could not go to school: 'We had nothing, I didn't have enough money to buy them a pair of shoes so how could I afford educating them?' Keejan then went on to explain that when her children were young there was inflation and poverty: 'Do you know what happened in Iraq? No one cared about the other then.' Her children worked from an early age. She feels angry because the Anfal surviving children 'are not accepted in any of the offices and organisations ... They [employers] say our children are uneducated.' The government's lack of interest in tackling this issue and passing the blame onto the children themselves for being uneducated has baffled Keejan. She is surprised by this because, Keejan argues, the children would have had an education 'if we were not Anfalised, if our husbands had not disappeared.'

Another obstacle to children's education was that before 1991 the children did not have Iraqi ID cards and because of this they were not accepted in schools. The women also lacked marriage certificates and ID cards so they had no proof that

---

19 Asmar, March 2006.

the children were theirs. After 1991, they found some legal loops and tried to get the children ID cards as their maternal or paternal uncle's children. This, however, led to other problems when the uncles decided to take charge of the children and deprived the woman of access to them. In some cases the uncle claimed the husbands' inheritance (see section on Social Problems below). Some children, however, because they had missed three years of schooling (1988-1991) refused to go back to school because they had fallen behind and they would have to study with students who were much younger than them.

'Successful' Mothers

Women who had fewer children, and no disabled child at home, or dependant elderly relatives, made sure that their children got an education. Khursha[20] who had a son and a daughter 'worked hard and suffered a lot' just to make sure that her children went to school. She was beaming with pride when she talked about her son, who is now a policeman, and her daughter who is a teacher: 'My life was ruined for lack of education and Anfal, I just wanted to make sure that I am the only one whose future is ruined. I didn't want my children to experience the same disadvantages, only my life should fall to pieces not theirs.' She also feels a great sense of satisfaction that her hard work has paid off and her children are both successful citizens with respectable jobs.

Rezan also managed to put her daughter and son through schooling. Unlike most Anfal surviving women Rezan had studied up to 2nd year secondary school. Due to her affiliation with one of the major political parties she benefits from more support than most of the Anfal women survivors. She works in the party's Woman's Union, receives the Anfal salary, has been given a house and her daughter receives support from Kurdistan Save the Children. All of these have helped Rezan in looking after her children. When talking to Rezan you realise that she is strong and content in her life. It is true that she has lost her husband and she has suffered through Anfal but the circumstances in the aftermath were more favourable in her case than most. Also, working in a woman's organisation and being an activist gives her power and satisfaction. She is defending not only her own rights but the rights of other women survivors.

Sending their children to school is reflected in both Khursha and Rezan's sense of self worth and self esteem. They feel that despite the extreme conditions, the violence, loss and poverty, they have managed to protect their children from experiencing further disadvantages. This has only been possible because both women have few children and they had no other dependants. Unlike the other women, these two women were not over-burdened with responsibility.

In summary, the role of mothers as the only breadwinner and care-giver to their children requires extra energy, draining the women and causing stress and exhaustion. This is particularly difficult for women who have large numbers of

---

20   Khursha, November 2005.

children, including some disabled children, and elderly dependants (their own parents or parents-in-law). Women who have been forced to work and leave their children at home alone, who have not been able to support their children's schooling and who generally feel that they have been absent and sometimes harsh, feel that they have been bad mothers. They feel responsible for the barriers their children face in society because of a lack of education and because of their cultural background. They blame themselves for a situation that did not involve any choice on their part (Belle and Doucet 2003: 104).

Exploitation in the Community

The women and their children were desperate to work and provided a cheap labour force in the community. Their illiteracy, or basic literacy, was a major obstacle to finding good employment. It meant that the survivors did all the low paid, low status jobs which were labour intensive and exhausting. Being poor and defenceless meant that they were vulnerable to exploitation and abuse. Here I will identify the ways that some women were exploited at work, sexually abused, and taken advantage of by the Kurdish media.

The majority of the women did not talk about being exploited by their employers or other members of the community. Most of them were grateful to be given work and to have an income even it if was too little. They lived in communities with other women who were in a similar situation. Many did not feel entitled to complain because, they believed, everyone was dealing with the same difficulties. It was after 2003, when witnessing the sudden rise of certain wealthy individuals, that survivors felt there was no equality and no justice. Most of them, however, felt that they could not betray their employers, landlords, landowners, and other individuals who had given them work and accommodation. Only a few women talked about being abused by employers who paid them very little and worked them too hard.

Various women reported that they 'worked hard for little money,' that they 'worked like men,' that they 'worked from dawn till dusk.' They worked 'every day' even when they were ill.

Nasik talked about some employers pushing them to work harder and not letting them 'straighten [their] backs' for a rest when planting or harvesting in the field. When they were picking cotton the employer stood behind them making sure that they did not miss out anything. Usually the women were given an hour's break at midday. Some days, however, if urgent harvesting was required, the break was reduced to 20 minutes.

While living in Bahirka, Ghazala reported picking watermelons, cucumbers and tomatoes and packing them. The women carried the boxes on their shoulders and took them to the storage a few hundred yards away. Each day they were paid 5 dinars for their labour which they spent on food for their young children. In the early to mid 1990s, there were plenty of women who were willing to do men's heavy work for little money. In retrospect some women are surprised that they managed through these difficult times. They also feel angry that their hard work, which contributed to the rebuilding of economy in the country, is not respected or valued.

Sexual Abuse

It is common knowledge that some women were exposed to sexual violence in the aftermath of Anfal but they refuse to talk about their personal experience of such issues (see Chapter 2, section on Sexual Abuse). Such experiences are only heard about second hand, through various rumours and whispers. A young woman worked in a man's house who gave her drugs and molested her. The man was later arrested by the Kurdish security forces. The young woman's mother decided to leave that town and move elsewhere because people were talking. Her sons were being bullied in school: 'Your sister was raped,' they were told, as if they were responsible for the injustice that had befallen them.

I first found out about this story through an interview conducted by a woman's organisation who pursued the family to their new place of residence. The mother was put on the spot in her workplace and was obviously not consulted before the interview. The woman tried, at first, to deny that her daughter had been sexually abused. But under pressure from the interviewer, who knew the details of the story, she started crying, covering her face and talking about it. She was begging them not to broadcast the story because 'honour is precious.' Unfortunately, this woman, who had been victimised by political violence and experienced violation in her own community was being victimised once again by a woman's organisation who wanted to 'get the truth out of her' at any cost. No respect was shown towards her choice not to speak, she was pressurised into 'confessing.'[21]

Other women talked about beautiful females being handpicked for work when other women, older and less attractive, were left out.[22] In the labourers square in a small Kurdish town women, who were dressed top to toe in black, would sit and wait at five in the morning for the land-owners to come and give them work. Some of the land-owners gave work only to the beautiful and young women. The attractive girls were isolated from their relatives and sometimes harassed in the field by their employers. Soon the women found a way around the system. Relatives stuck together and told the landowners they would only work in groups. 'For the sake of each beautiful young girl two other women got a job.'[23]

---

21   This conversation took place between the woman's officer and myself after I watched the interview in their centre.

22   Adalet Omar, Erbil, March 2006.

23   Conversation with Mahabad, a woman's worker in Khatuzeen centre, Erbil, March 2006.

Media Abuse

The media also abused these women. Journalists in search of making sensational documentaries regularly called on the survivors and interviewed them about their plight. The women were expected to cry and lament in front of the camera. When women tried to draw attention to their current problems most of the time this part of their story was edited out (see also Chapter 6). Some women who still lived in Iraqi controlled regions were urged to talk about Anfal experiences, leaving them without protection when the programs were broadcast. A young woman and her mother who lived in Duz Khurmatoo, which was under Iraqi control until 2003, had been interviewed before the fall of Saddam's regime. The journalist reassured the two women that he was conducting the interview for the archives and that it would not be broadcast. Despite this reassurance, the interviews were soon shown and the two women had to go into hiding, in fear of being arrested. Understandably, the two women do not trust anyone with a camera. When I went to see them in 2006, they were really angry and they refused to talk to me.

In summary, there is evidence that women all over the world 'work longer hours for lower wages, often in sweatshop conditions, and [they] may be severely isolated' (Bassuk et al. 2003: 51). Also necessity and hardship mean that women may be 'powerless in their dealings with employers, landlords, and government bureaucracies' (Belle and Doucet 2003: 104). Women feel angry and disappointed when members of the community fail them. Many reported that they understand why Arab soldiers and officers may beat, hurt, and abuse them but when Kurds do it, this is experienced with more pain. The fact that their labour, plight and stories have been used for various purposes by different agents: employers, journalists, politicians, researchers, women's organisations and other individuals, has caused a great rift between these women and the rest of the community. They struggle alone with mental and physical health problems.

Social Problems

One of the biggest problems of the Anfal families is in fact social problems ... We need a cultural revolution to solve these problems. (Adalet Omar, Erbil, March 2006)

'A man is the umbrella on one's head,' says the Kurdish saying, meaning a man is a woman's protector. Having no man (or no father in the case of young women) means that a woman is exposed to social and economical problems. In this patriarchal society, where honour is closely tied to a woman's body and where there are strict concepts of 'honour' and 'shame' a woman who has no man, as her protector and patriarch, is always viewed with suspicion. As a result other male members of the family, brothers, brothers-in-law, uncles and other male relatives, try to control the woman and closely watch over her and her children. Traditionally it is considered 'a matter of honour'[24] for a man to take responsibility for the girls and women in the family. Most of the time, however, this does not translate into caring for a woman, financially or emotionally, but it implies controlling and abusing her.

To have a husband or a father means that there is one person, one man, who is making decisions and taking charge. Usually, this man cares about the woman (and his daughters) and even if he is authoritarian he would be loving, he would provide for them, and sometimes he may even negotiate things with his wife and daughters. To not have such a man means that the door opens for many others to interfere in the woman's life. Suddenly many men see themselves as the rightful guardian and they would be making decisions on behalf of the woman and her children. They take charge and believe that they are protecting their own honour and that of the whole family and tribe. Gullalla clearly pointed to this when she talked about her brothers-in-law taking charge. She said: 'If a woman goes under a men's authority then she loses control over her life.'

Anfal surviving women and girls experienced many social problems in the absence of a 'rightful' male guardian. This was especially true if the woman did not have a grown up son, or brother, who would defend her against his uncles and cousins. The young girls were particularly victimised in this process. Their uncles and cousins had the power to prevent them from going to school, and marry them off for money or advantage without consultation. Some women were forced to marry their own cousins, or be exchanged as a wife for male relatives. They were prevented from marrying a man of their choice and prevented from leaving an abusive marriage. Young women who rebel against these restrictions risk further stigmatisation and control. In some cases they risk losing their lives (see below).

The women themselves were oppressed in various ways. Some in-laws forcibly deprived mothers of their children and told her to leave the house. In some cases young and childless widows were prevented from remarrying. Widows who had children risked losing access to them if they chose to remarry. Some women were abused by their brothers and brothers-in-law. Their income was taken from them, and they were beaten, harassed, and abused. Women who ended up living with other relatives were sometimes treated like servants. They had to work hard in the house, look after nephews and nieces, and tend to the farm. Some of them were not even provided with enough clothing and basic needs. They were unloved and uncared for.

Sometimes their husband's families were angry with them for surviving when their own sons had not. There were also problems with inheritance. Most of the women did not have legal marriage certificates as they had married in small villages and the ceremonies were conducted unofficially by an Imam. After Anfal they, and their children, were deprived of what was rightfully theirs (I will explore this further below).

---

24   Yusif Yunis Jawhar, Duz Khurmatu, December 2005.

Losing Children

Awaz had 4 girls and 3 boys. After losing her husband and surviving Dibs camp she was living in the same courtyard as her brother-in-law in Smood. The brother-in-law did not allow the girls to go to school. He believed that girls would eventually get married and a man would look after them so, according to him, they did not need an education. He also wanted Awaz to stay at home and look after the children, especially as the girls started to grow up. Awaz defied him because if she did not work her children 'fell asleep crying for food.' She carried on working for four years until he 'threw [her] out.' Awaz recounted that her brother-in-law kept the children and made her leave 'in the clothes [she] was wearing.'

Awaz was told that she was unwanted: 'Our son has disappeared, what are you doing here? Go get remarried.' She went to Kifri to stay with her brother and she was unhappy there: 'I became a servant to my brother and sister-in-law and they still did not appreciate me.' She was desperate for her children, crying every day, and thinking that she will go 'crazy.' Her younger brother kept telling her not to cry. 'How could I not cry?' she told him, 'I raised my children with so much difficulty and now they have taken all of them from me (she cries). I am left alone, how could I not cry?' She felt that 'death is better than this.' Her brother went and persuaded the other family to let the children come and see her but they allowed only the boys to come. After this the youngest boy did not go back and he stayed with his mother.

Eventually Awaz registered a complaint against her brother-in-law. They brought the children to court. Their uncle had 'threatened the children' and pressurised them to say that none of them wanted to live with her. Only two of the boys dared to speak out and they stated that they wanted to stay with their mother. Over the years the brother-in-law gave her daughters away for marriage without consulting her or the girls. The married daughters now come with their husbands to visit Awaz. The girl who is not yet married is not allowed to come to her mother. They claim that it is the girl's choice, that she does not want to see her mother but after what she heard from the other children Awaz is convinced that her daughter does not dare.

Another woman, who was suspected of being raped in prison, was 'thrown out' by her parents-in-law and deprived of seeing her four children (see section on sexual abuse). Her in-laws condemned Ahoo for 'speaking with Arab soldiers in prison' and they were angry that she had survived while their four sons had not. Ahoo returned to her father's house and she is grateful that he 'took [her] in.' She was not able to see her children for six months which made her 'go crazy.' After a long process of negotiations, which involved a lot of begging and crying on her part, her father-in-law took her back. But after a few months they 'threw [her] out again.' They told her to go and get remarried because the children don't belong to her anymore. Ahoo was tearful when she remembered how her children were crying and 'holding on to [her] legs' when they forced her to leave the house.

For a few years, this woman was only able to see her children at the neighbours', who felt sorry for her and secretly brought them to her. After a while a number of people intervened on her behalf and talked to her in-laws. They persuaded them to at least let her see her children every now and then. Since then she has been allowed to see them sporadically. But 'there are still restrictions,' Ahoo tells me. The older child, who is now married, comes to see her regularly. In a manner similar to Awaz's case, it is still more difficult for the unmarried daughter to be with her mother.

This woman also had to fight for her children's love. The youngest one in particular was influenced by her grandparents. She believed that her mother was dead: 'My mum didn't listen to my dad and my dad threw her into the well.' For a few years the young girl believed that this woman was her aunt. But the in-laws did not succeed in 'brainwashing' the older children. Now, after all those years, as this woman is getting older, struggling with health problems and depression her parents-in-law 'are even regretful for what they did to [her].' Despite everything she is forgiving. She believes that a part of the problem may have been that they could not feed her even though she herself worked and sometimes she secretly gave her children money.

Honour Killing

Saza was only eight years old when Anfal reached her region. Her father was Anfalised and she, along with her mother, two sisters and a baby brother, moved back to Taqtaq to live with her grandparents. As a young teenager Saza fell in love with the boy next door and she was seen speaking to him in the neighbourhood. Her mother's uncle heard about these rumours and one day he came and packed their basic furniture into a van, without consulting anyone. He moved them to his own place where he gave them a room in the courtyard. When the old grandfather asked him what he was doing, the uncle angrily told him that it was none of his business.

Two months later, at the age of 14, Saza was given away for marriage without her knowledge. She only found out that she was getting married when the religious ceremony, nikah, had already been concluded in her absence. Soon she was given away and her husband took her away to his far away town. Years passed, and Saza gave birth to two sons and adopted a baby girl who had been 'thrown away.' Saza recounted that her husband regularly beat her up and kicked her out of the house. Once, he beat her and told her that he did not like her. In fact he was only keeping her because of the boys. He then hired a taxi and told the driver to take her back to her brother's house.

It was during this taxi journey back to her hometown that she met her previous boyfriend again. She was sitting in the back of the car, crying for herself and for her children. The car had stopped in heavy traffic in Taqtaq and he, who worked as traffic police, recognised her and started knocking on the window. She did not dare to open the window but he kept begging her to talk to him. He had looked for her everywhere but no one had told him about her whereabouts. The man followed the car and found out where she was staying and started making contact with her, declaring that she was the love of his life. This is when Saza, who could not shake off her abusive husband (he kept coming back for her and taking her back by force), started a relationship with this man.

I interviewed Saza in March 2006 when she was hiding in a woman's shelter in Suleimanya. Her husband and relatives had found out about her affair and wanted to kill her. Saza believed that all of what happened was because her father was Anfalised. If her father was alive, she believed, he would let her marry the man she loved. He would not force her into marriage with a man she didn't know, and he would protect her from abuse. On 20 December 2007 Saza was shot in the head by her brother. After staying for a while in the woman's shelter, which resembles a prison, Saza was fed up. She missed her children, and she agreed to make peace with her husband and go back to him. Her family promised not to kill her and she promised to give up her affair. Unfortunately this ended in her death.[25]

Saza's murder is an example of 'honour' killing, a practice which affects not only the Kurdish community but South Asian, Middle Eastern and North African communities amongst others (Distcheid 2003). Kulwicki (2002: 77, 82) defines this to be 'crime committed against women by their male family members because the women had violated the honour of their family ... [by] alleged sexual misconduct.' Women may be killed to clear the family's name. A woman is said to bring shame on her family if she commits adultery, if she has a sexual relationship prior to marriage, or if a widowed/divorced woman becomes pregnant. In some cases women are killed for less severe reasons, such as refraining from an arranged marriage, or refusing to be exchanged with another bride. She may be killed because her refusal is interpreted to mean she is having a relationship with someone else. It is inevitable that other women survivors have been victimised by 'honour' killing for various reasons. Their stories and their voices, however, are missing from this research and the mainstream Kurdish history and discourse.

The Role of Women's Organisations

In the post liberation era, after 1991, various organisations were established to defend and protect women's rights in Kurdistan. Some of the women who started these organisations were inexperienced. Others lacked understanding and awareness. It was 'necessity' that made some into women's workers.[26] As a result some of them inadvertently made mistakes and they were heavily criticised because of this. The majority, however, have made great progress in protecting women and securing more rights. Various women's organisations tried to help Anfal surviving

25 Rewan, No. 162, 16 January 2008.

26 Conversation with Runak Faraj, The women's media centre, Suleimanya, December 2005.

women by defending them in court, supporting their claims, and trying to give them control back. Belle and Doucet (2003: 109) point out that 'advocacy efforts to help women resolve legal problems, gain the benefits to which they are entitled, or secure education can be therapeutic in themselves.' This is because they give the individual a sense of not being alone in this world, they provide hope, and renew faith in humanity.

Ikrama Ghaeb,[27] who is the director of the Women's Centre in Sarqalla, has dealt with a few such cases. Two sisters, whose father disappeared during Anfal, were forcibly married off to two brothers, their own cousins. One of the girls seemed to be getting on well with her husband while the other was really unhappy. She wanted to get divorced but was beaten by her husband and her cousins (his brothers). She kept running away from her husband and she was forced to go back. Ikrama went first to the senior politician in the region whom she knew and who was supportive to her work. After securing his support she then filed a complaint against the man and he was forced to divorce his wife. He was also issued with an injunction forbidding him to go near her. The girl eventually remarried and she seems to be doing well.

Other times, however, solving these problems is more complicated especially if the woman has a deep sense of abiding by tradition, even when this is oppressive and humiliating. There is a legal way to deal with disputes regarding inheritance, custody of children, forced marriage, and honour killing but tradition still plays an important role in the community. The structure of the society limits women's reactions and choices. Jack (1991) suggested that women influenced by cultural norms and gender-roles, adopt a gender-specific schema called 'silencing the self' which influences their choices of behaviour in intimate relationships.

Women may silence their thoughts and emotions and exhibit self-sacrificing behaviour in an effort to maintain an intimate relationship. Silencing the self has been found to be related to depression in women (Jack 1991). Jack suggested women become depressed 'not by the loss of a relationship, but by the recognition that they have lost themselves in trying to establish an intimacy that was never attained' (Jack 1991: 27). Jack and Dill (1992) found that the schema is activated most often in interpersonal contexts wherein women's needs and freedom of expression are devalued. It develops within a society that 'discounts femininity itself – its knowledge, its perspectives, its values' (Jack 1991: 33).

Abiding by tradition and silencing one's thoughts and desires is typical of women in the whole community but may be particularly strong in rural areas where tradition is stronger. It affects the choices of not only the young widows who may want to remarry but also unmarried girls. They may choose not to marry because they are looking after disabled and elderly family members. Sometimes, not marrying, not having joy in life, is considered a sign of loyalty to their perished fathers, brothers and cousins. Grieved mothers are more likely to scold their daughters for having fun than their sons: 'How can you have fun

---

27   Ikrama Ghaeb, Kalar, May 2009.

after what happened to your father?' Mothers take pride in their sons wanting to get married yet if a girl states that she would like to marry they will be considered 'shameless.'

Also, young women who have missed out on education and do not have a chance to go out and work, are more likely to be isolated and undiscovered. Their chances of meeting men and finding a suitable match are severely limited. As a result of lack of opportunity, social sanctioning on women's lives, and the fact that many men became victims in the various wars, there are large numbers of women who have never married. Not having a husband means being deprived of all the things that only come in the context of marriage in this society such as: love, sex, children, home, and independence.

Another factor which influences women's choices is also inherent in the structure of society. Women may refuse to go to the courts and claim their rights because they are scared of further violence and pain, especially if the man is politically well connected or wealthy. Taban Abdulla pointed to this issue when she talked about encouraging women to go to court and claim custody over their children. The women refused because 'she could face violence.' If a woman cannot come forward and ask for help, no one can interfere in her family affairs. Women's officers are working in a difficult environment and they have to be careful not to push women to do things which may endanger their life: 'You have to always guard the woman so that she does not get killed by the relatives of her husband.'[28]

The way around this has been to negotiate with people and to continue the dialogue in the hope of persuading them to do what is right and fair. Taban pointed out that: 'We have been able to pressurise the man's family, in a few cases, socially pressurise them to give the woman and her children their due. But of course we could not help everyone in Kurdistan.' This also points to the limitations of NGOs and women's organisations who cannot tackle these complex issues alone. These organisations, by their own nature, have limited resources to deal with the structural inequalities in society. Most of their resources are dedicated to advocating individual women whose lives are in danger, supporting women's cases in court for custody over children, for inheritance, and the right to divorce, and women's entitlement to support and housing. Some women's organisations have also worked on changing awareness in the community by providing various programmes and initiatives aimed at young people,[29] providing training to women themselves about their rights and entitlements,[30] however these programmes are limited in scope and need more backing by the government and by other International NGOs.

There is widespread disapproval about the women's movement, headed by the women's organisations, all over the region. Even though some of the criticisms

---

28   Taban Abdullah, Suleimanya, November 2006.

29   Khatuzeen's workshops with schoolchildren about democracy, women's rights, honour, shame.

30   Kurdistan Women's Union, Centre for Media and Education of Women, New Life for Anfal Women.

directed at women's organisations may be justified,[31] overall this thinking is clouded because it fails to see that the responsibility of securing better rights for women, poor people, disables people, and the minorities does not lie with NGOs alone. A major responsibility lies with the government. Blas et al. (2008) point out that civil society aims can be restrained by governments that are unwilling or unable to achieve greater equality. The state, therefore, must support the civil society through providing a regulatory framework which includes:

> (1) recognition of the political legitimacy of civil society and a community's voice; (2) involvement of civil society in all its forms in policy development, implementation, and monitoring; (3) ratification and implementation of legal protection for civil society organisations; (4) design of policies that transfer real power to people; (5) resourcing of policy implementation to support community empowerment; and (6) reform of professional education to give greater status to lay and indigenous knowledge. (Blas et al. 2008: 1688)

The government must respond to the NGOs' pressures and suggestions. There is need for a holistic approach to protect women's rights, prosecute deviant men, and support single women by providing financial support, appropriate housing, training and education opportunities. Activists and women's officers need to be very careful because in certain cases it is possible to 'use threats' and tell a man that he will be arrested if he does not do what is right but in other cases they have to tread more carefully. There are still people who can evade justice and avoid punishment because of their tribal or political allegiances.

Services Provided by the Kurdish Government

In 1998, the PUK administration in Suleimanya and Garmian, and in 2000 the KDP administration in Erbil and Duhok started providing a salary to support the families of 'the martyrs.' This was a vague term used to refer to those who had died during armed conflict or had been killed because of their affiliation with the Kurdish revolution. Some Anfal survivors went forward to register their disappeared members of families as martyrs. This, at the time, was frowned upon by some individuals and institutions because, they believed, it would lead to a

---

31   The majority of the women's organisations started as the women's organ of various political parties. This posed huge obstacles to collaboration between these organisations because of political differences. Things have changed, however, as increasing numbers of independent women's organisations have come into existence. Various forms of collaboration have taken effect in the last ten years.

formal recognition of those who disappeared during Anfal as martyrs which in turn would mean reduction in the number of Anfal victims.[32]

Over the years there have been various governmental institutions and ministries that have dealt with Anfal survivors' needs. In the PUK region, for example, at first the institution was called the Martyrs Foundation. In 2001, this foundation became part of The Ministry of Social Affairs. In 2004, the foundation became an independent ministry, called the Ministry of Human Rights, the Displaced and Anfals. In 2006, after the creation of the joint PUK-KDP administration both foundations were joined up under the name of The Ministry of Martyrs and Anfals Affairs which had directorates in the three main governorates of Suleimanya, Erbil and Duhok.[33] Over the years the various institutions issued a number of decrees regarding the rights of the families of the Anfal victims. Only in 2007, with the establishment of the Ministry of Martyrs and Anfals Affairs, was there a clearly defined law regarding the rights of Anfal survivors.[34]

According to Law Number 37 of 2007 the Ministry of Martyrs and Anfal Affairs made distinctions between those who were 'Anfalised' and those who were 'Martyred.' According to Article 1 'Anfalised' people are those who 'have lost their lives or have been harmed or their fates are unknown as a result of crimes of genocide against the people of Kurdistan.' Martyrs, on the other hand, are divided into two groups: 'Martyr of Struggle' which refers to those who 'have lost their life during armed conflict or political struggle to defend the liberation movement of Kurdistan against the successive repressive regimes or as a result of the mass killings or war acts against the people of Kurdistan,' and 'Citizen Martyr' which means 'anyone who is martyred as a result of war acts or acts of terrorism or during the mass exodus of the Kurdish people.' These definitions in turn affect the first grade relatives' entitlement to government benefits. Martyrs of struggle are paid the largest sum, followed by Anfal survivors with survivors of gas attacks, and finally Citizen martyrs.

In the second article of the same law it is stated that this law is to secure rights and provide benefits to the relatives of martyrs and Anfals, respecting and supporting them 'in a way that is consistent with the weight and position of the martyr's sacrifices to reduce the pains of their relatives.' This ranking approach to

---

32 Taban Abdullah, Suleimanya, November 2006. This was confirmed by Adalat Omar who also confirmed that in Erbil there was resistance to recognising Anfal victims as martyrs (conversation with Adalat Omar, August 2009).

33 Establishing this was rather difficult as various people provided me with different information. Finally as a result of various conversations with Adalat Omar, Abdulkarim Haladni, Saman Qader and Ali Bandi I was able to establish this chronology. There is a general sense of chaos when you go to the offices and institutions because staff keep changing and new staff are not usually aware of the department's history. Old staff tend to move on and don't remember details. Official documents are not easily accessible.

34 Conversation witt Abdulkarim Haladini, The Suleimanya Directorate of Martyrs and Anfals Affairs, August 2009.

those who have lost their lives as a result of various episodes of political violence has angered many survivors. Behe who complained that the Kurdish government is rich now and could provide better support to the Anfal families stated: 'Let them consider us martyr's families, at least the martyrs have graves, the families have proof of what happened to them, they are treated better, we don't even have that.'

From the beginning there were a number of problems with the support provided. First of all, when support for survivors started, salaries were minimal and did not distinguish between a family who consisted of two survivors and an extended family with young and/or disabled children and elderly dependants.[35] This led to family conflicts in some cases when a widow and elderly parents-in-law fought over, and sometimes had to share the already minimal salary. This changed recently with the new law, when both a disappeared person's parents and their spouse are equally supported.[36] Secondly, those who had contact in the government or were supporters of one party or another were more readily supported. Although there have been attempts to eradicate this problem it is not completely resolved (see Khursha's complaint below). Thirdly, the amount received was the same whether a family had lost one member or ten. Now the salary is double for those families who have lost more than 4 members to Anfal and it doubles again if they have lost more than 9 people.

Most importantly, however, entitlement to support did not automatically mean that women had access to support. Generally speaking the process of getting help was particularly difficult for them. Even reaching government officials to seek help can be difficult for single women who live in remote towns and villages. Article 3 of the same law states that those who are to fall under this law must provide 'official documentation to prove their positions.' If they do not have the necessary official documentation 'they have to prove through other legal channels before the committees, the court, or specified sides and then put before the General Directorates in the provinces of the Region.' The burden of proof here lies with the survivor, not the government. Many of the women do not have marriage certificates. Proving that a disappeared man is their husband required providing two witnesses and a long bureaucratic process involving a lot of paperwork. Some women had difficulty claiming their husband's salary or his inheritance. The same goes for children who have died in the camps or during the gassing, shelling, and

---

35   In 1998, at the time of the Martyrs Foundation in the PUK region, survivors were awarded a monthly income of 150 Iraqi dinars which was equivalent to $15. In 2002 this was raised to 200 dinars ($20). With the change of Iraqi currency in January 2004, and the increased wealth of the regional government due to the political changes, the salary was raised to 90,000 new Iraqi dinars ($60) and then in 2005 to 150,000 ($100). More recently the salary was raised again to 450,000 dinars ($380) for those who have lost up to three members of the family to Anfal and 900,000 ($760) for those who have lost more than three.

36   Conversation with Abdulkarim Haladini, Suleimanya Directorate of Anfal and Martyrs Affairs, August 2009.

flight. Women needed to first prove that they were married, that these children were actually theirs and that they had actually died in the process. These barriers and the bureaucratic system baffled women and made them even angrier towards the Kurdistan government.

WHO (2009: 16) points out that the position of individuals in the socio economic hierarchy shapes access to resources and every aspect of experience in the home, neighbourhood and workplace. Some dimensions of an individual's socio economic position are determined by 'education, income, occupation, and prestige.' The low socio economic position of Anfal survivors affects their access to support. Speaking of the difficulties of access in the context of the American Benefits System Belle and Doucet (2003: 108) point out that government efforts to tackle poverty should take into account measures that ensure actual access to benefits that poor people are entitled to. Barriers to receiving available help instigate more anger amongst survivors. They feel that they are battling against a large bureaucratic system which treats them with suspicion.

Various women complained that they feel they are not respected when they turn to governmental bodies and offices to seek support. Keejan argued that 'Now when we go to an office no one even respects us or turns to us. And then they [Kurdish politicians and media] shout about Anfal. What is the point? If I cannot do anything for my children then I would not be shouting about them all the time.'[37] Khursha said:

> We served peshmarga for many years and now when you go to an office, they
> don't even look at us ... I am not saying they have not done anything for us but
> it is not enough. We go to them as Kurds. They should not discriminate between
> KDP, PUK, The Communist Party, etc. It was not just the Kurdish revolution
> that was attacked, we as Kurds were attacked. (Khursha, November 2005)

The support that is given is seen to be irreconcilably short of the extent of loss and suffering experienced by survivors. Keejan referred to this when she said: 'I have lost four people in Anfal, I lost my home and land and property. We have no village left.'

The rise in salary has created a further social problem for women who are now expected to stay at home. Taban Abdulla pointed to this when she said people are telling survivors: 'Now that you have a salary, why don't you stay at home and look after your children?'[38] The salary, however, does not cover rent, water, electricity, health bills, food, clothing and furniture. It needs to be supplemented by women's work. Semeera has four daughters, two of whom suffer from severe epilepsy and have large health bills. She complained that if she did not work as a cleaner in the primary school her children would not manage. According to this woman the monthly salary 'is not enough for 15 days.'

---

37 Keejan, December 2005.
38 Taban Abdulla, Suleimanya, November 2006.

The government also provided some basic housing and land to the survivors. According to Saman Qader of the Ministry of Anfals and Martyrs Affairs, the Kurdish government first started distributing land in 1999 and housing was provided from 2002. In total 25,886 pieces of land and 5,011 houses have been provided in the three governerates of Erbil, Suleimanya and Duhok. The question remains, however, as to who has received these services and who has been deprived of them? Ruqia who lost a son and a daughter to Anfal is very angry at the Kurdish government because she did not receive the housing she is entitled to: 'The government (KRG) has given houses to jash but not to me ... they divided houses here and they gave them to jash, those who killed peshmarga, those who are rich, and I live amongst scorpions.'

Razaw argued that whatever happened to them (Anfal victims and survivors) was part of the struggle for independence. The Kurdish government was established because of the sacrifices of the Anfals and martyrs. It was Anfal that gave the Kurds a profile in the world, according to Razaw: 'Did anyone know who the Kurds were before Anfal?' Yet she believes that their sacrifices are not recognised by those in power, that the Kurdish government has repaid people by neglect. She, along with many other survivors, feels that no one cares. She talked about her husband's aunt who is the only survivor of her own family. She has lost her seven children and all her grandchildren to Anfal and now she is homeless. The government has provided some housing for Anfal survivors but her husband's aunt was told that she does not qualify because she is on her own. The old woman now lives with her niece. She is a restless soul who keeps visiting her various relatives every day but she can never rest anywhere, she has no peace anymore. 'How could they do that to her while she has lost all her family in Anfal?' Razaw asks, 'The government should look after her.'

The services provided by the Kurdish government are not coping with the demands for support, compensation and housing. This is aggravated by injustices when previous members of the jash forces, collaborators with the Iraqi government who fought the peshmarga, because of their tribal and political connections are more readily supported, even though they are less in need. Survivors are exhausted and depressed by the poverty and the discrimination they have experienced for many years. Now they are expressing their demands in a more angry fashion. It was probably this angry crowd who voted for the 'Change List' on 25 July 2009 securing the new list 25 seats in the Kurdish parliament.[39] The unsatisfied survivors regularly come on this group's TV expressing their dissatisfaction with the status quo.

---

39   Listi Goran, led by Newshirwan Mistafa (a close friend and former comrade of Jalal Talabani) put forward a new list comprising of capable administrators, writers, academics who because of their lack of political affiliation with the KDP and PUK have been sidelined. In its first election the list secured 25 seats out of 110, putting pressure on the KDP-PUK coalition government to address corruption and invest in infrastructure.

Loss of the Farming Community

> Why should I let them live there like donkeys who don't know anything? For the
> wheat? I don't want their wheat. We've been importing wheat for the last twenty
> years. Let's increase that for another five years. (Ali Hassan Al-Majeed, Meeting
> with members of Northern Bureau and governors of Iraqi Kurdistan, April 15 1988,
> cited in MEW 1993: 347)

Anfal crippled agricultural production in Kurdistan. The campaign not only
destroyed the majority of the farmers and deported the rest, it also involved
burning down the farms, orchards, and vineyards; destroying the water springs
and canals; and looting animals and farming machinery. Initially, after extensive
reconstruction projects in the rural areas throughout the 1990s, Kurdish farming
experienced a short period of revival. Despite the absence of agricultural machinery
and limited man-power the villagers started farming on a small scale. A number
of factors contributed to the long term failure of these attempts. These include a
shortage of agricultural labour, urbanisation, drought, civil war, the UN's Oil for
Food programme, and lack of sufficient support by the Kurdish government.

The large decline in the number of men during Anfal meant that the majority
of the remaining labour force consisted of women, children and the elderly. The
shortage of reliable agricultural labour made it difficult to embark on agricultural
activities on a large scale. Yet despite this and despite a lack of tractors, water
distribution channels, and animals the villagers produced fruit, vegetables, wheat
and barley, and dairy products because there was a market for these goods. The
sanctions on Iraq and Kurdistan and restrictions on imports meant that home
grown products became popular once again.[40] However the gap in farming years,
between 1988 and the mid-1990s, lead to loss of skills. The new generation, Anfal
surviving children, spent some crucial years of their lives in towns and cities,
doing odd jobs. They were unfamiliar with farming and some of them showed
no interest in carrying on with their parents' and grandparents' professions. They
migrated back to the cities.

The rapid and forced urbanisation caused by Anfal, the gap years between the
village destruction and the opportunity to return, and the traumas associated with
returning (see Chapter 6) meant that some families chose not to return or they
returned for a short period and then left for the cities once again. There were better
living conditions in the towns and cities. This included the availability of basic
daily services such as water, electricity, sewage system, hospitals, schools, and
better security. People returned to the towns and cities because of their children's
schooling, the availability of better job opportunities that were less energy
consuming, and a lack of infrastructure in the villages – bad road conditions
leading to isolation, limited number of schools, in particular secondary schools,

---

40   Ruqia, November 2005.

lack of hospitals and health centres, and minimal basic services such as water and electricity.

The region has also experienced several stretches of drought over the last ten years. The village of Kulajo, for example, consisted of 20 households after its reconstruction in the mid-1990s. Now only five families live there. Two families have left the village in the last three years because of the drought. Insufficient rain meant a lack of pasture for animals. Those who used to live on animal farming moved to the cities. Sebri who returned to Kulajo in the late 1990s, left in 2007. She explained that lack of rain meant there was no food for their animals. She went on to say that her children preferred life in town because it was more comfortable.

The civil wars between the various Kurdish parties have also contributed to this problem. The village of Warte was on a fault line between the joint KDP-PUK forces on the one hand and the PKK (Kurdistan Workers' Party, Turkey) on the other. The situation deteriorated rapidly as people in the village, who were supporters of one of the Iraqi Kurdish parties, got involved in this battle against the PKK. According to Qadir[41] even when peace was restored between the PKK and the Iraqi Kurdish parties, the war between the Warte villagers and the PKK still continued. This made the area unsafe and farming became difficult. A similar problem affected Koreme village which also became a war zone between KDP and PKK. In other areas, the war between KDP and PUK, PUK and the Islamic movement, and both sides with the PKK affected general security and made it unsafe for villagers to continue their work.

Despite its corruption[42] the UN's Oil for Food programme (between 20 March 1997 and 21 November 2003) gave Iraq the opportunity to sell oil, authorised by the UN Security Council, to fund the humanitarian program which provided the Iraqi public with basic food stuff and medicine. The program was later expanded to include 'infrastructure rehabilitation and activities in 24 sectors: food, food-handling, health, nutrition, electricity, agriculture and irrigation, education, transport and telecommunications, water and sanitation, housing, settlement rehabilitation (internally displaced persons – IDPs), mine action, special allocation for especially vulnerable groups, and oil industry spare parts and equipment.'[43] There was, however, a lot of criticism about this program which was perceived as centralised and political. It had no intention to help the region thrive but was merely interested in keeping the status quo and preventing death by hunger. According to Shamal Abul Waffa, the previous Minister of Agriculture in the PUK region, many corrupt individuals and companies benefitted from the program.[44]

---

41   Qadir, March 2006.
42   See MacAskill (2005) and Lynch (2005).
43   United Nations (2009).
44   Shamal Abul Waffa, previous PUK Minister of Agriculture, Suleimanya, November 2006.

Throughout the duration of this program about $31 billion worth of humanitarian supplies and equipment were delivered to Iraq.[45] The food, however, was brought from other third world countries and was delivered to Iraqi people as part of their food rations. The provision of free flour, rice, oil, soap, and other dry goods meant that there was no market for locally produced rice and wheat anymore. More recently, lifting the Iraqi sanctions on 21 November 2003 meant that the door was opened to the import of goods from abroad. Being an oil producing country meant that it was cheaper to import than to produce (Segal 2009: 19-20). The locally produced fruit, vegetables and dry food cannot compare in price with the cheap imported goods. Therefore agriculture has become less and less viable.

Another factor which was mentioned by women is lack of support from the Kurdish government. According to Shamal Abul Wafa,[46] 'Agriculture was not a priority of the Kurdish government and there was no strategic plan to revive it.' Some of the problems, he argued, were related to lack of a holistic approach to this issue. The Ministry of Agriculture alone could not solve everything because 'some of these problems are related to lack of infrastructure in villages' and this is to do with other ministries' work. The Kurdish government was dealing with a destroyed terrain which needed rebuilding, there were high rates of unemployment, interference by regional powers into the Kurdistan region, and it was also bogged down in its own internal conflicts. In the chaos of a divided administration and because of corruption and a failure to get priorities right, agriculture took the back seat.

For the revival of agriculture it is necessary that a number of issues are holistically tackled. These are: good infrastructure in the villages, including roads, schools, health centres; providing logistic support such as water, agricultural machinery, good seeds, pesticides, etc.; protecting home produce by providing a market, applying quality control, and imposing restrictions on imported goods; supporting and compensating farmers at times of crises such as drought; and lending money to farmers who are keen to restart agriculture and those who want to improve their produce.[47] The majority of the population in Kurdistan now is completely dependent on cheap imports of food and rations provided by the government. This made it unnecessary for the villagers to grow their own food. Also the dependence of farmers on government handouts, lack of government support for agriculture, and various political and social obstacles have prevented the revival of agriculture in the region.

45   United Nations (2003).

46   Shamal Abul Wafa, previous PUK minister of Agriculture, Suleimanya, November 2006.

47   Shamal Abul-Waffa, November 2006.

Conclusion

The majority of the women who survived Anfal were farmers who worked in their own farms, orchards, and vineyards and tended to sheep, goats, cows, and poultry. They produced wheat, barley, rice, chick peas, lentils, fruit and vegetables, dairy products, and meat. Anfal deprived them of their homes and their means of production. They suddenly became dependant on others for survival and they had to start their lives from scratch, most of them as single mothers. The characteristic feature of life after Anfal was the continuous struggle to find a suitable home and employment. Some women are still struggling to find appropriate accommodation even though things have gradually improved because of help provided by the Kurdish government.

Before Anfal these women had worked alongside their husbands on the farms but they did not have control over the income that was generated from their work. They were also less likely to make decisions within the family about daily life and social activities. After Anfal the women became the sole breadwinners and guardians in their families. Most of the women reported putting on men's trousers (Kurdish baggy trousers) and working 'like men.' They built houses, carried heavy loads, and laboured on other people's land. Despite all of this the community continued to perceive these women as 'zaifa,' a term generally used to refer to women and literary meaning weak. This in turn is reflected in some women's perception of themselves.

In the immediate aftermath of Anfal the majority of the women were poor and they were desperate for work. Their situation was exploited by various individuals such as employers, landlords, the media, and others. Despite their minimal education and basic skills, they were resourceful in finding various avenues for work. Some women are now unwell as a result of years of intensive labour. Some had no choice but to revert to illegal and socially unacceptable work which caused stigma. Poverty also influenced women's self-image. They blamed themselves for going out to work and leaving their children at home, for the fact that they could not afford to educate them, and for the current disadvantages their children suffer from in society.

Women were stigmatised for not having a male guardian and the stigma was transferred onto their children as well. The lack of a 'suitable' male guardian meant that various male members of the family and tribe assumed responsibility for the women and made decisions on their behalf. There is evidence that poverty, inequality and discrimination put women at high risk for depression (Belle and Doucet 2003: 109). Addressing depression in such circumstances implies combating the structural inequalities, including women's social inferiority in relation to men, that influence women's daily experiences and put them at a disadvantage.

Social exploitation also took place when women, who had no Identity Cards and marriage certificates, were deprived of their children and their husbands' inheritance. In some cases women were killed by their own families for going out to work, for alleged sexual misconduct, or for simply refusing an arranged marriage or wanting to get out of one. Various women's organisations tried to help defend Anfal surviving women. Their work, however, was restricted by the tribal structure of society and by how influential the defendants were. Those who have major support and influence in one of the Kurdish parties are likely to escape punishment.

In the last ten years the Kurdistan Regional Government has provided a number of services to Anfal survivors including a monthly salary, housing, and land. Only in 2007, however, a law was passed clearly stating the rights of the survivors and their entitlements. The new law attempts to amend some of the injustices of the past, as pointed out by survivors. Nevertheless there are still problems regarding access to support and also the amount of support survivors are entitled to. Women complained that what they received as a way of salary was not enough, that their specific conditions (such as having disabled children) has not been taken into consideration and that despite all their sacrifice previous members of the jash are more readily supported by the government.

Farming never recovered since Anfal. Despite short periods of hope, the farming community is disappearing. Environmental, political, and economical factors have contributed to this loss. Lack of investment in the rural areas went hand in hand with political instability during the civil war, droughts, and importing cheap goods from abroad. The UN Oil for Food program also contributed to this because it provided free food stuff to Iraqi civilians in return for oil. The various factors have meant that there is no market for local goods and it is too expensive to produce. As a result farming seems to be finished in Kurdistan even though there are no industries and other means of production.

Chapter 6
The Psychological Consequences of Mass Violence

Anfal was a violent process of destruction and rupture. It involved physical harm and injury, and mental harm through witnessing destruction and death. The campaign deprived the village population of their homes, their families and tribes, their farms and means of employment, and their places of worship. The majority of the women survivors are poor: they live in sub-standard housing in villages and housing complexes with minimal services. Their society perceives them as hopeless widows and mere victims and these are unhealthy reference points for identity and self image.

Addressing the wellbeing of women survivors of Anfal requires recognising the links between the social, psychological and the physical domains of life. It requires acknowledging a past full of injustice, violence, and loss as well as tackling their current socio-economic status. This is consistent with what Watters (2001) describes as: 'a move from Cartesian dualism towards a more holistic approach incorporating mind and body' (Watters 2001: 1713).

According to WHO (2000) Mental health is:

> ... the capacity of the *individual*, the *group* and the *environment* to interact with one another in ways that *promote subjective well-being*, the optimal development of individual and collective goals consistent with *justice* and the attainment and preservation of conditions of fundamental *equality*.

This definition is useful when addressing mental health consequences of Anfal because, for the survivor, it is their relationships with their community and their political, social and physical environment that have been disrupted. During the Anfal campaign, the political environment became increasingly hostile until it culminated in mass murder and genocide. The betrayal of Kurdish collaborators who deceived people under various guises and persuaded them to surrender, and the sense of abandonment by the Kurdish revolutionaries who could not protect the civilians, and in some cases even prevented their escape, caused great rifts between survivors and their own community.

Anfal involved great injustices and led to the creation of a deprived and isolated underclass. It enhanced the social and political inequalities already inherent in Kurdish society. The ad-hoc housing complexes built by the Iraqi government for the deported communities have now become small unattractive towns with minimal trees, paved roads, and no entertainment facilities for young people.

Poverty, marginalization, and a lack of voice have caused great anger amongst the survivors and in particular amongst their children. This is heightened by the knowledge that previous collaborators with the Iraqi government have secured high positions within the Kurdish political milieu; that they are wealthy, powerful and well represented in the government.

The above definition, therefore, draws attention to the network of relationships that determine mental health and acknowledges the importance of social context, justice and equality. Gender impacts on mental health at all the levels of individual, group and environment and particularly in the existing differences in the attainment of justice and equality (WHO 2000, Williams 2004, Solomon 2003: 5).

Loss and Trauma

It is widely acknowledged that losing valuable resources such as family, possessions and social networks can place mental health at risk (Green 2003: 17). Such losses are the defining features of the lives of Anfal surviving women. They have also been relentlessly exposed to a range of events associated with 'actual or threatened death, serious injury, or sexual or other physical assault'.[1] These events may cause reactions such as intense fear, horror, numbness, and helplessness (Solomon 2003: 7). The psychological reactions in turn can have consequences for the wellbeing of survivors.

The two most common disorders associated with torture and mass violence are Post Traumatic Stress Disorder (PTSD) and Major Depressive Disorder (MDD) (Turner et al. 2003: 191). PTSD is characterised by three sets of symptoms. Hyper-arousal when the person has a heightened arousal response making them easily startled. Intrusion when they have recurrent flashbacks and nightmares about the traumatic event. Finally, there is numbing when the person avoids thinking about the trauma or anything that reminds them of the trauma (Herman 1997). There is disagreement as to how common PTSD and MDD are amongst trauma survivors. Summerfield (1999: 1449) disputes the view that all survivors of violence are traumatised. The author stresses that PTSD is a 'reframing of the understandable suffering of a war as a technical problem to which short term technical solutions like counselling are applicable.'

Silove et al (2000) argue that disagreement in the field seems to arise from this 'either or' fallacy. Some scholars suggest that entire populations need psychiatric help and thus they medicalise a normal response to extreme violence while others argue that change can only be achieved at a socio-political level, which stands to assist the whole community. Silove et al (2000) argue that both extremes are mistaken. The authors draw attention to the need to identify separate sub-populations with distinctive needs.

---

1   http://www.istss.org/resources/what_is_traumatic_stress.cfm.

It is arguable whether counselling and therapy are conducive to the mental health of Anfal survivors. First of all, this individualistic approach is not in accordance with the communal identities and communal ways of copying used by survivors. Secondly, survivors suffer not merely because of their past experience with violence and loss but also because of their current low socio-economic status. Providing therapy to survivors without addressing their daily struggle with poverty, sexism, stigmatisation, and lack of acknowledgement is to ignore a large part of the problem. Such programs can only be useful when they are synchronised with community development programs that provide survivors with opportunity and practical help.

This approach is at the heart of feminist therapy. According to Brown (1994) feminist therapy aims at developing feminist consciousness. This means to help survivors develop an awareness that 'one's own suffering arises not from individual deficits but rather from the ways in which one has been systemically invalidated, excluded, and silenced because of one's status as a member of a non-dominant group in the culture' (Brown 2004: 464). The ultimate goal is hence to empower survivors by providing 'strategies and solutions advancing feminist resistance, transformation, and social change in daily personal life, and in relationships with the social, emotional, and political environment' (Brown 1994: 21-22). In this sense feminist therapists are not content with 'simply healing the individual clients in front of them' but they want to bring about 'political action ... to end interpersonal violence, end political oppression and repression, and work for peace at both micro and macro levels' (Brown 2004: 470). This has been supported by others who have confirmed that political action to bring about change, whether it is to combat sexism, racism, classism, or any other form of discrimination, is usually experienced to be therapeutic and healing (Aron 1992, Comas-Diaz 2000). I will explore these issues below.

Survivors' Psychological Reactions to Anfal

Generally speaking there is stigma around mental health problems in Kurdistan. People are perceived as sane or they are not. Those who suffer from hallucinations or who talk to themselves arc considered 'mad people,' and until very recently there was no recognition of the variety of ways that distress can be expressed. Lately as a result of the work of mental health NGOs, and the broadcast of mental health programmes on Kurdish satellite channels the issue has found space in the public domain. Some of the terms used in Western mental health discourse have no comparable terms in Kurdistan. For example, the term 'trauma,' used to be translated as 'hedma' which was taken from the Arabic word 'sadma.' The word literally means shock. Recently 'trauma' has been translated as 'zebir' literally meaning 'force.'[2]

---

2 Heartland Alliance, Kurdistan, 2008.

The Kurdish community is fed up with the survivors' painful stories. Once, when I spoke about my research to a woman she said: 'The Anfals?' and then, in a mocking manner, she put on the dialect of a woman from Garmian while she said, 'I have no father left, no mother left, no one left. That is all they have to say.' In the spring of 2007 I provided a series of seminars to young people in Kurdistan about Anfal surviving women.[3] I was shocked to hear what some of the students had to say about Anfal. In one case a young man stood up and stated: 'We are fed up with hearing about Anfal, we should try to forget it.' There was a general lack of interest in what I had to say. Some people believed they knew everything and there was nothing new to be learnt about Anfal. Herman (1997: 115) addresses this lack of interest in, and hostility towards survivors when she says:

> The chronically abused person's apparent helplessness and passivity, her entrapment in the past, her intractable depression and somatic complaints, and smouldering anger often frustrate the people closest to her.

It would not be true to say that Anfal surviving women are trapped in the past, that they are helpless and passive. Understandably, most women experience periods of helplessness, passivity, anger, and depression. Most of the time, however, they are getting on with their lives as best they can. Nonetheless, this unfair image of 'eternally helpless and depressed survivors' is widely presented in the Kurdish media which provides an incomplete and skewed picture of the survivors (see section on voice, below).

In the interviews I conducted no woman referred to herself as traumatised. Yet Anfal survivors show a range of psychological reactions such as fear, numbness, insomnia, depression, rage, desire for revenge, guilt, shame, and personality change. Some of these were experienced immediately after the traumatic experiences while others are more long term reactions to Anfal. Below I will first explore some of the psychological reactions reported by the informants and then go on to talk about ways of coping with these experiences.

Fear and Guilt

Traumatised individuals may experience intense fear in the immediate aftermath. They may fear that it is happening to them again. Herman (1997: 86) points out how the sound of sirens, slamming doors, and thunder may cause terror in traumatised people. They feel that something terrible is about to happen. This was Srwa's[4] experience. She was 14 years old during Anfal. She survived when her seven siblings and her niece, all of whom were with her, perished. A member of the jash forces took pity on them and told them: 'I will save two of you for your

---

3   This was facilitated by The Ministry of Sports and Youth which organised the seminars in the Universities and the Youth Culture Houses.

4   Srwa, December 2005.

parents' sake. You must decide who will stay behind.' Despite her older sister's insistence that they should stick together Srwa decided to stay behind with her 9 year old brother. When the IFA truck came to take her siblings away her brother could not handle it. He ran out to the soldiers and begged them to take him: 'I am their brother too,' he kept telling the soldiers, 'I am one of them.' Srwa was biting her hand as she watched from a window while the soldiers loaded her siblings into the trucks that took them to their death.

The Kurdish collaborator hid Srwa in his house until she found the remainder of her family: her parents and two sisters. Though reunited with them, while the family tried to escape to Iran, she was terrified that the soldiers would come for them. Some nights she jumped out of sleep, screaming from fear. Her family did not feel safe to travel for a while because they were worried that she would scream and give them away. The fear of Anfal did not leave Srwa for a number of years. She is known for having suffered from severe mental health disturbances such as extreme fear, screaming, depression, feeling confused, and experiencing fits of rage. People who referred me to her told me that she had 'gone mad' for a while.

In Srwa's mind the day of their capture has acquired mystic qualities. Thousands of people were steered towards Milla Sura where they were met by the army. Srwa remembered the confused crowd, the helicopters circling above, the soldiers shouting orders in Arabic, the IFA trucks that were being loaded with people. To her 14 April 1988 resembles no other day: 'It was spring and a lovely season but you felt as if the sky was raining blood. When you looked at the sun it felt as if a huge crime would be committed on that day.'

Srwa still feels the terror of that day. She reports looking into her siblings' faces and feeling that she would never see them again. To this day she wonders to herself why they didn't turn back. They saw thousands of people heading towards Milla Sura. They saw the helicopters. They felt that something terrible was going to happen but they did not turn back. As a child she had heard of 'black days' and 14 April is the embodiment of such a day: 'Every year on that day ... the world is foggy, it is windy, it is always terrible.'

Srwa still feels alienated amongst happy crowds. Her older sister tries to take her out to picnics and parties but she usually ends up crying. Happy occasions seem to take her back to the past. While her sister sympathises with her, she knows that she cannot share Srwa's feelings because it was Srwa who watched on as 'a large section of [her] family [went] to their death.'[5] Srwa feels angry with herself because she did not raise her voice when they were taken away and because she did not go with them. She chose to live but she now feels guilty about her choice. She suffers from what Langer (1991: 122) calls tainted memory, 'stained by the disapproval of the witness's own present moral sensibility,' after living in a 'more privileged time' when she could afford the luxury of a moral system. Similarly, in a study with Holocaust survivors 15 to 20 years after incarceration Matussek et al. (1975: 23) found that many of the survivors were 'plagued by feelings of guilt' for

---

5 Srwa's sister, November 2006.

surviving even though they had no power to protect anyone. Perhaps, as Garwood (2002: 362) points out, guilt is a reaction against the utter helplessness survivors experienced at the time of the atrocities, it is a desperate attempt to regain control by taking responsibility.

After various visits to a doctor, complaining about muscular pain and depression, Srwa was advised to do something she enjoyed. She started writing poetry but one day when she was grappling with her past experiences she collected all her poems and set fire to them. It all felt pointless and she felt that poetry 'could not solve her problem'.[6] Normal enjoyable activities felt futile to Srwa as a survivor. After what happened to her family during Anfal she is not sure whether words meant anything anymore. Some days nothing makes her feel better. Darkness descends on her which she cannot fight.

Numbness and Flashbacks

Another immediate reaction may be feeling numb which later makes the survivor feel guilty about her lack of 'appropriate emotional response.' Nian reported feeling numb during her journey to safety. She was amongst the group (including Nazdar, below) who managed to escape to Iran during the first Anfal attack when most of the people who fled with her froze to death. Two days earlier the peshmarga had tried to open a route through the snow but a vicious snowstorm blocked the mountain routes. She walked, carrying her one year old daughter in her arm and she kept walking as people slipped and fell, no one stopping to help them get back on their feet. She was so battered by the snow and cold that she did not realise it when she lost her shoes on the way. The snow clumps kept hitting her and she only realised that her own face must have 'turned blue' when she looked at other people's faces.

Nian recalled, in great length, her journey through the snowstorm when her one year old daughter nearly froze to death, her retreat from the mountain, and her escape via another route. For her the difficulties started after reaching Iran because when Anfal was in full swing there was 'no time to think and reflect.' Later she felt guilty for her lack of emotions in response to the unfolding catastrophe: 'How could I not feel anything? When my daughter nearly died ... How could I not have any reaction?' Only later she grasped the extent of her loss and the tragedy. The inability to mourn as the catastrophe is unfolding is typical of other survivors of violence as they are 'involved in their own life and death struggle' (Garwood 2002: 358). Guilt is usually experienced in the aftermath when the survivor assumes some form of responsibility in the face of paralysing forces (see also Chapter 2).

In the months that followed Nian experienced personality change. After settling down in her new home in exile she regularly felt as if she was 'in another world,' that she was 'a different person.' While living in Iran she was 'haunted by the images and the memories.' She thought about her house, her friends, some of

---

6   Srwa's sister, November 2006.

whom had frozen to death, their forced displacement from the 'liberated zone,' the collapse of the Kurdish revolution, and the loss of a way of life. She found it difficult to put her memories to rest: 'I can never forget about them, I can never leave those moments behind … I have told dozens of people about these things. I would never forget these scenes.'

Nian also seems to have suffered from PTSD immediately after Anfal, as the images of death and destruction, people falling around her and not getting up, people giving up and lying down to die, kept being replayed in her head. These images 'made a home in [her] brain,' and she felt that her 'soul was decayed by the catastrophe.' For many months she felt hopeless and felt that she 'could not do anything for [herself].' She 'could not escape these thoughts.'

Traumatic Pain

Nazdar[7] grappled with intense fear and pain for a long time. She and her sister, Khuncha, were 18 and 14 years old respectively when they tried to escape to Iran via Kanitoo mountain (see Nian's journey, above). Having witnessed large numbers of people freezing to death the sisters were terrified in the aftermath. At twilight while retreating from the mountain they had walked past corpses of acquaintances and friends. In one case a man had died while standing stiffly. He was up to his chest in the snow. His eyes were open and Nazdar had gone close to him, thinking that he was still alive. 'We should take him with us,' she told her father. He reassured her that somebody would help him out but they should keep walking as 'the government is close.' He did not want to scare her by confirming that he was dead. Later Nazdar had regular flashbacks. 'We did not have the courage to go anywhere alone,' recounted Nazdar, 'We had seen so many dead bodies that we thought they would come back to us.'

When I asked her whether she was scared Nazdar replied: 'I was very very scared. For about a year I did not dare to go out in the night. Whenever a door slammed I thought there was a plane above my head ... If I came out at night I would run back inside and think that the dead were following me.' Sudden loud noises reminded Nazdar of planes and missiles and the darkness brought back flashbacks of the dead. Nazdar and her sister lived with these fears for a long time. Some nights her sister, who was four years younger than Nazdar, would jump out of sleep and call for her: 'Nazdar, I can see the dead, I feel that they are all around us.'

The two sisters' experienced hyper-arousal and intrusion symptoms which are components of PTSD. During the hyper-arousal stage the survivor 'startles easily, reacts irritably to small provocations, and sleeps poorly' (Herman 1997: 35). Intrusion, on the other hand, takes place when survivors 'relive the events as though it were continually recurring in the present' and these flashbacks are triggered by small reminders.

---

7   Nazdar, June 2010.

Nazdar's hands and feet were completely numb in the snow. She could not hold anything and she had to be spoon-fed for what seemed like a long time after her escape. She still experiences the same pain every now and then. Four years ago, her hands became so painful that she could not do any housework. 'These three fingers were so painful I had no peace,' said Nazdar, 'The children were still young then and I kept thinking how am I going to wash the dishes and clothes?' She then went to the doctor hoping to find a cure.

The doctor could not find any objective grounds for Nazdar's pain. He explained that all those years ago the blood in her hands must have frozen and that must be the reason why she now has pain. He sent her on her way after prescribing some pain-killers. 'Some days I used to cry from pain and I worried that my hands would be cut off. I asked the doctor whether I had an incurable disease (referring to cancer). He told me that I didn't. He said he would give me some medication which wouldn't cure this but it would help.'

The pain usually recurs in cold weathers. One spring day she went along with other women to pick kingir (a root vegetable) from the fields. A rain storm started and Nazdar was shivering once again. She returned home with difficulty, her legs were weak under her: 'I become useless in the cold, I can't even carry a bucket of water.' Each time she goes out in the cold she has to come and sit by the fire: 'The children keep asking me why I like the fire so much? I say it is because my hands and feet froze in the snow.'

Twenty years after Anfal cold winter days have the capacity to revive the terrible memories for Nazdar and she experiences the same pain again in her hands and feet. Her experiences are similar to those of other survivors who re-experience the same bodily sensations which they had at the time of exposure to extreme trauma (Kardiner and Spiegel, cited in Herman 1997: 45). This is because traumatic memories are encoded in the form of 'vivid sensations and images' which intrude the survivor's consciousness (Herman 1997: 38). It is possible, therefore, that Nazdar's pain is a remaining part of her memory intrusions, a component of PTSD.

Grief

Some survivors cannot let go of traumatic memories involving the loss of a significant other such as spouse, parent or sibling. For some this results in 'a series of personal problems in which unassimilated grief ... is central'. (Veer 1998: 16)

Nasreen[8] seems to be still grieving for her husband, brothers-in-law, and the other men who were shot on the same day in her village, Koreme. The women and children (inhabitants of Koreme and Chelke villages) had been separated from their men on site. Thirty-two men were lined up and shot just before their families were loaded into the military trucks which took them to prison. The men were later buried in four collective graves in the place of the shooting, amongst the almond trees to the east of the village.

---

8   Nasreen, June 2010.

In 1992, nearly four years after the incident, when a team of forensic experts uncovered the Koreme graves[9] Nasreen found the bodies of her husband and one of her brothers-in-law (the other one had been arrested and disappeared without a trace). She was there when the bones of the men were uncovered, recognised only by their clothes. The images of their broken and pockmarked bones are etched into her memory. She remembers men's shirts 'filled with bullets' but the image she found most painful was that of her husband's skull which was dotted with bullets 'like a paper pierced with cigarette burns.'

Her husband's suffering and the pain he must have felt at the moment of shooting is something which Nasreen cannot escape. 'I will never forget,' Nasreen says while crying, 'may no one see what we saw.' She then goes on to explain that till the day she dies: 'I will talk, walk, laugh but the pain of seeing them like that will never leave my heart.' Immediately after the graves were unearthed Nasreen fell ill and was hospitalised for a couple of weeks.

Nasreen's grief is fresh and strong after 22 years. This may be particularly so because she had six young children and no grown up brother, brother-in-law or son who could help her in the aftermath. She worked hard and lived in abject poverty. She still lives in someone else's house in the same village and is struggling to make ends meet. These conditions have accentuated her low spirits, preventing her from letting go of those painful images. She has thought about and imagined the men's pain in detail, how they were shot under the oppressive August sun:

> Their bodies were left in the sun for a long time. The soldiers had then set fire to the fields and my brother-in-law's leg was burnt up to the knee. Then they were thrown on top of each other in the pits ... This is an immense grief. May no one experience the pain and poverty we experienced.

Depression and Isolation

The inability to escape moments of devastation makes some of these women feel isolated and detached from the rest of their families and community. Some of them may be unable to enjoy or maintain relationships with loved ones (Green 2003: 23). This is particularly difficult because family and community support is one of the major protective factors that enable women to survive trauma and hardship (see section below). Keejan reported staying awake many nights, while her children sleep, 'until the day cleans the night.' She sometimes watches her children as they 'come and go, they sleep, walk, sit.' They are trying to get on with their lives and Keejan tells me that 'they were young, they have forgotten.'

Keejan is scared that their carefree existence may not continue: 'I just stay awake and think about all the lost things. Our lives were ruined, it is finished. God

---

9   A team of forensic experts from Physicians for Human Rights uncovered the Koreme graves in 1992. The findings were then published in a joint report with Human Rights Watch: Middle East Watch and Physicians for Human Rights (1993).

have mercy on my children.' She feels lonely in her fears and her fixation with what has been. She still dresses top to toe in black and she cries when she talks about Anfal, 'as if it happened yesterday.' Like many others, Keejan reports going to wakes where she can cry for the lost life and relatives.

Keejan lost fourteen male members of her extended family including her husband, brothers in law and cousins. She also lost three children in the camps where she was detained for four months. The last child was a six year old girl who suffered from epileptic seizure. The child craved for cucumber in her last days. The day she lost her life she chewed on a green slipper on her mother's lap thinking it was a cucumber. Eighteen years later the smell of cucumber still gave Keejan 'a headache.' It is a constant reminder of her daughter's last cries, her hunger, deprivation, and suffering.

Contaminated Memory

For many women the basic, simple things have been contaminated with memories of violence which gives them the power to trigger flashbacks. A hospital worker[10] told me that one morning he had bought a loaf of bread and he offered a piece of it to the cleaner, a woman survivor of Anfal, who started crying. 'How could I have warm bread?' she told him, 'My two year old son wept for bread till he lost his soul.' The boy had died in Nugra Salman camp, the fort on the border of Saudi Arabia.

Women reported the intrusion of Anfal memories into their daily lives. Meena[11] regularly remembers a baby boy, who was separated from his mother in Topzawa camp. 'I will never forget that,' Meena says, 'I will never forget, never.' The mother, who looked older than she was, was separated from the younger women to join the elderly. She kept turning back, pointing to her son and repeating 'My baby, the cradle.' The soldiers told her she would return in a few minutes but she never came back and the boy was left there in the courtyard. Meena often wonders about what happened to the little boy, she cries as she tells me about him. She often remembers him when she sees a cradle, when she hugs and touches her own children.

Meena spent a large part of her detention time in Nugra Salman camp where they were locked up in the halls at night. Even now the sound of chains makes Meena jump, it reminds her of nights when the guards chained the prison gates. She also hates closed doors and is reported to keep the doors open even when she goes to the toilet: 'I never close doors. Even when I visit people I ask them not to close the doors, it saddens my heart.' Closed doors remind her of prison days.

Razaw[12] talked about Anfal as a wound that never heals. She argued that every time the survivors talk about it, hear about it, or see TV programs about it their

---

10   March 2009.
11   Meena, May 2010.
12   Razaw, May 2010.

wounds are picked once more: 'It is an injury without recovery.' She believes that no one who has lived through the process can forget that pain 'until we die, until they put us in the ground.' She had been a teenager when Anfal took place and she survived the camps while her sister and 6 nephews and nieces perished. Twenty-two years later she is married and has children of her own but she can never forget Anfal. Many days at sundown she feels that her heart is 'being squeezed' because it was at this time when they took her relatives away. On daily basis she is reminded of Anfal. On happy occasions if she sees someone who looks like one of her missing relatives she feels the pain again.

Revisiting happy times can also be painful. Keejan says that she cannot endure going back to Bangol village where she lived with her husband and eight children amongst the extended family. Visiting the place where she first had a family, where her husband lived and worked is too much to bear for her. She believes that going back there will drive her 'mad.' She is sensitive to reminders of her lost life.

Women who have moved back to their villages, report that locations take them back to those days when they harvested the fields with their disappeared sisters, brothers, and husbands. They remember their orchards, their old houses, the peaceful days of togetherness. Jwana[13] talked about the field where she worked with her siblings, cousins and friends. 'We used to harvest together and sing,' Jwana recalled, 'We were so happy then. Now when I go to that field I want to cry (she bursts into tears). I keep thinking that is my brother's place.' When her nephew, who is the only survivor of her sister's family, comes back from the US to visit, she watches him and cries: 'He looks like my sister Sarah,' she says, 'I get sad when I see him. He keeps telling me: Don't be sad, what does sorrow do? They have gone now, he says, don't kill yourself grieving for them.' The village itself, now inhabited by five families only, feels like a place full of ghosts. The ruins of some of the houses are still there, a reminder of the lives that were lost.

Embodied Distress

In her study with Cambodian women Herbst (1992) found that exposure to severe traumatic events led to somatic symptoms, as well as memory and attention problems. Similarly, many witnesses in this study reported physical complaints. The most common complaints were back and shoulder pain, headaches and numbness of the head, general lack of energy and feeling lethargic. Some of this may be related to the hard labour over many years which has taken its toll on their health but it is also likely to be embodied distress arising from their past experiences of violence and rupture.

It was also common to hear from the women that they were experiencing memory and concentration problems. Habiba's[14] husband disappeared in Anfal and she lost her toddler son in Nugra Salman. She was regularly muddled up

13   Jwana, May 2010.
14   Habiba, April 2006.

about the chronology of events, the number of the children who accompanied them (her own children and her brother in law's), and the age of her children. Even though she is only in her early fifties she seemed to confuse different periods in her life. She kept apologising for her bad memory and she related them to her Anfal experiences: 'We have seen so many awful things we can't remember things well.' After asking her the same questions to clarify the matters she told me that 'this needs patience and time to recall everything' because they have 'suffered a lot.' Her recollections are sporadic.

Anger

After the initial shock, hopelessness and depression many survivors reported being enraged at what they have been put through and especially angry with perpetrators and collaborators (Veer 1998: 18). Women felt angry watching Saddam Hussein on TV, talking from the luxury of a warm court room, living in a proper cell and being fed three times a day. They felt angry that he was allowed to have a trial, a right which was not granted to any of his victims. Keejan is baffled that Saddam Hussein sits opposite a Kurdish judge and is allowed to defend himself: 'How could he defend himself? ... all those beautiful children, all the young men, all the villages, what did he leave?' She reported that every time she sees him on TV her blood pressure rises.

Some women argued that they wanted Saddam Hussein to be hanged in Kurdistan where they could go and watch. Others demanded that he should be handed over to the Anfal surviving families. Behe said: 'We request that Saddam Hussein should be killed, but they should do it publicly. Why do they keep him and sniff him like basil, showing him to us on TV?' Other women, however, were aware that his death would not bring them peace of mind. Gelawej[15] told me while touching her heart and talking about her lost youth and her shattered life: 'Nothing can make this feel better, nothing.'

The witnesses also feel angry with the Kurdistan government because the burden of the Kurdish revolution fell on the shoulders of the villagers who had to feed the peshmarga on daily basis even though they themselves were poor farmers. Then, when Anfal took place it was mostly the civilians who were captured and disappeared. The failure of the Kurdish parties to protect the civilians and in some cases, preventing their escape, has lead to a lot of resentment. Finally, the survivors feel angry because the Kurdistan Regional Government has not done enough to support and involve the survivors in the post liberation era, and because their claims for justice, for putting the jash leaders on trial, have been ignored.

---

15   Gelawej, December 2006.

Resolution

Another major issue for the survivors was lack of closure that characterised their lives over the past two decades, which contributed to feelings of depression and helplessness. After the end of violence, which could not be avoided but could only be endured, women ended up waiting for the return of their disappeared husbands, fathers, sons, brothers and sisters. Despite reports published by Human Rights Watch and captured Iraqi documents, which talked about mass killings during the Anfal campaign, they continued to hope that their loved ones were still alive. Many women reported that they were hopeful until the fall of the Baath regime in 2003. At this stage when the political prisoners were released and especially when a number of mass grave sites were identified, it became evident that those who disappeared during Anfal were not alive anymore. Yet the process of suspense continued because the majority of the mass graves have not been uncovered due to the unstable political situation and ongoing violence in Iraq.

Many women in this research talked about their need for closure. They want the bodies of their loved ones to be recovered. They consistently obsess about where their loved ones have ended up, how they have been killed, where they are buried. The fact that the Anfal victims have had 'no wake, no fatiha (Quran recital for the dead), no coffin, no grave'[16] makes it difficult for women to move on. Srwa pointed to the importance of having a grave. She said that when you have a dead body you become 'certain', you know that it is 'finished.' However, because Anfal victims don't have graves, Srwa argued, each time the survivors talk about them 'it feels as if it has happened right at that moment.' Srwa believes that not having a grave is not 'in accordance with nature.' Even death during Anfal, she said, was not in accordance with nature because some of the victims were alive when they were covered with dirt and living human beings should not be buried in graves.

A woman's officer who is Srwa's sister, also pointed to the importance of certainty. She argued that her own waiting is to do with not having closure. She knows it is irrational, that her siblings are almost certainly dead but she cannot stop hoping: 'It is all related to Saddam Hussein and his trial, if he tells us how he destroyed them, how he sold them. If he ever admits...' she reported that every time she hears that a Kurdish woman has been sold to an Arab country (see Chapter 2, section on Sexual Abuse), she hopes it may be one of her sisters. Every time she hears about a little boy being rescued by an Arab family, she hopes it is her brother. From her work with other Anfal survivors she pointed out that they all felt the same: 'When someone dies, you go to their grave, you become certain, you cry for them and remember them. But we don't know which grave we should cry on. We don't know.'

Ghazala[17] lost her husband and two sons during Anfal. Her husband was amongst the men who were shot in Koreme. Her teenage sons were separated from

---

16   Behe, November 2005.
17   Ghazala, June 2010.

her in Nizarka camp. The body of her husband, who was buried in a mass grave in Koreme, was recovered in 1992 (at the same time as Nasreen's and Nalia's husbands). The body of her sons, however, have not been found. Ghazala kept wondering how her sons were killed and where they were buried. She wished they too had a grave like their father, a place she could take comfort in visiting. This type of anguish may be experienced more keenly by women because of their family responsibilities.

In the spring of 1992 Nasreen, Nalia, Ghazala and many other women and men from the villages of Koreme and Chelke walked back and forth to the gravesite every day for about ten days until the process of uncovering the bodies was completed. Until she found the body of her husband in one of the graves Nalia had hoped that he was still alive. Five men had escaped from this incident and this gave the women hope. Nalia was glad that the team came because even though the five survivors informed her that her husband had been shot dead, she still could not believe it. She maintained hope for a miracle.

During the months and years before uncovering these graves Nalia was all ears every time someone knocked on the door. She kept hoping that her husband would come back, or that someone would bring news from him. His body was found in the fourth and final grave, along with eight other men. When they found him she touched his shirt and announced that it was him 'with his milky-coloured Kurdish uniform, his khaki shirt, and his shoes which were white.' They put the bodies in small boxes, 'as big as a pillow.' These were big men's bodies fitted into small boxes. She took the box which contained her husband's bones and went under an almond tree. She thought: 'That is it, it is over. These are his bones. I will never see him again.' She was finally certain that he was dead and she was able to cry and lament. Afterwards the 27 boxes were brought back to Koreme, each box covered with the Kurdish flag on top of a taxi. Every taxi was full of the relatives of each victim. They did the Islamic prayer, read the Fatiha, and buried them in Koreme according to tradition.

The need to know and to be certain is vital for the wellbeing of survivors (Quirk and Casco 1994). It helps people to accept the loss, to grief properly and to finally move on. Having lost hope about their relatives being alive and in the absence of closure women are left in a status of limbo. Some women choose to believe that a miracle may be possible and that their loved ones may still be alive somewhere. Most women experience daily swings between hope and hopelessness which exhausts them and prolongs the road to recovery.

Coping with Anfal

Like survivors of other cases of mass violence the Anfal survivors were displaced from their homes and had to cope with numerous losses, trauma, poverty, isolation and stigmatisation. Espin (1992) points out that loss does not only include the big things such as family, socio-economic status, support networks, and employment, but also includes more subtle things. She mentions the absence of familiar everyday smells, daily routines and generally lack of 'the average expectable environment' (Hartmann 1964, cited in Espin 1992) where the person knows where she is, is familiar with the environment and feels safe and secure. The absence of these little things, Espin (1992) maintains, become a constant reminder of being uprooted and these can be most disorienting and disruptive of the person's previously established identity.

The risk of mental ill-health is not related to exposure to stressors in a unilinear fashion, but may be affected by a range of personal and social factors. There is evidence that, for example, certain personality traits are more useful while coping with stressful life events. Ferguson (2001) found that conscientious and positive personalities are more likely to use active coping strategies and this is more conducive to mental wellbeing. This, however, does not take into account the extremeness of the stress encountered and the support that is accessible at the time (Saldaña 1992: 27). Exposure to extreme traumas may wear down the spirit and exhaust the strongest person's resources (Bassuk et al. 2003: 33).

When you meet Srwa, for example, it becomes clear that you are meeting an intelligent, eloquent and intense young woman who has survived a lot of pain and guilt. She had been a clever and beautiful teenage student but Anfal prevented her from completing her education or leading a normal life. Green (2003: 18) stresses that exposure to extreme traumatic events 'may create such disruption that the usual coping strategies do not work, and adjustment to them may not be possible.' Permanent damage to survivor's copying capabilities may result. It is important, therefore, while talking about survivors coping with Anfal the multiplicity of the factors that affect coping are addressed. Here I will identify some factors that may help survivors cope with Anfal.

Protective Factors

The factors that help the individual cope with stress are usually referred to as 'protective factors' (Ager 1993). There are a number of protective factors including social support and religious affiliation (Ager 1993, Watters 2002). There is evidence that social support may help individuals recover from violence and may generally be beneficial to mental health (Green 2003: 20, Ager 1993, Veer 1998). This is mainly because mass violence disrupts communities and destroys normal support networks, isolating the individual. Social support can work to sustain the individual through difficult times and provide her with a sense of belonging and security. Support can be emotional or instrumental. It can also be provided by different groups (Biggam and Power 1997). Each individual has a number of support networks at each time, some based on kin relationships, others on friendship and others still on important acquaintances. Here I will talk about instrumental support provided by family, community, and NGOs and go on to talk about the importance of religious affiliation.

Members of families, for those who had any left, were mentioned as a major source of support. After Anfal Srwa was encouraged by her family to get out of the house and do something. Her father and sister helped her find work in 1996, eight years after Anfal had come to an end. She worked in a governmental office in the Garmian region where many people have lost members of their families to Anfal. Some of her childhood friends worked in the same place. This helped Srwa a great deal because she was working in a supportive environment with people who could relate to her experience. Getting out of the house was a good means for Srwa to attempt to have a normal life. Her process of recovery, however, was disrupted a number of times due to the political instability of the region during the civil war between the Kurdish factions[18] and due to her own mental health situation.

Many survivors reported being helped by members of the community, many of them complete strangers, in the aftermath of Anfal. When the September General Amnesty was announced the villagers from Badinan were released from Nizarka and Salamya camps only to be deported to a far away desert outside Erbil. This was a hot flat plane far away from their native cool and mountainous region. Truckloads of people were brought to this area and dumped without a roof over their heads. A military base nearby kept watch on them. They had no income, no belongings and they had no freedom of movement. It was the people of Erbil who helped the thousands of civilians survive. Various survivors from different villages in Badinan reported how the civilians from Erbil came to Bahirka bringing carloads of blankets, food, tents, clothes and even doors and windows and bricks to build houses. Mullah Fatah Balakani, an Imam from Erbil, is reported to have advised those who wanted to go to pilgrimage that year that they should give their money to the displaced Kurds in Bahirka instead.[19] That, he had told them, was more virtuous than spending their money on going to Hajj.

Similarly many people from the region of Garmian who ended up in Smood housing complex (later renamed Rizgari, Liberation), Kalar, Kifri, and Sarqalla reported being helped by the inhabitants of these places. In the wake of the amnesty people gave the women money and food. Many opened their doors to the dispossessed people who came and stayed for various periods until they were able to reunite with other surviving relatives or find housing and means of employment. The group who were most desperate were the elderly who had returned from Nugra Salman camp having lost all young members of their families. Due to months of starvation diet in the camp, and thirst and illness, some of the elderly were not even capable of looking after themselves anymore. Some of them were taken in by strangers while others ended up living in dumps and people gave them food, water and sometimes a wash 'for God's sake.'[20] These moments of kindness may have helped heal some of the damage created by past trauma. However, the scale of

---

18   During the civil war, Srwa who was associated with one of the political parties, was later sacked when the other party took control of the region.

19   Ghazala, June 2010.

20   Ruqia, November 2005.

the catastrophe, loss, poverty, and grief was not comparable to the small remedies provided by members of the community.

Communities may also be a source of stress for individuals who go against tradition and 'they may interfere with coping and healing' (Green 2003: 21). Women who may have been raped are largely stigmatised in the community (see Chapter 2, section on Sexual Abuse). A more common case of community pressure, however, is directed at widows who have remarried. The majority of women survivors have not remarried by choice. Some women defend the memory of their dead husbands and argue that they loved their husbands, they had a good life together, and they do not want to 'live another life after him.'[21] The few women who have remarried, however, are considered unfaithful and untrustworthy. 'How could they?' is a question regularly asked about these women. How could they forget the memory of their disappeared husbands and, in some cases, leave their children with their in-laws, to restart a life with a new man? This is when men whose wives have disappeared have soon remarried without eliciting any negative reaction from the community. It is generally agreed that men need women to look after them, to feed them and wash their clothes and raise their children for them whereas women are expected to sacrifice their lives to the memory of their lost husbands and to looking after their children.

Various international and local NGOs as well as women's organisations set up initiatives to support the women survivors. Some of these initiatives failed[22] while many proved to be a success. In the post-1991 society many NGOs got involved in rebuilding the villages and helping survivors. Between 2001 and 2003 Norwegian People's Aid (NPA), for example, provided widows with five lambs each. The women were not allowed to sell the five sheep for five years. Each spring, in the lambing season, one lamb would be taken back from the woman to feed back into the project and keep it going.[23] In the Garmian region 287 women from 50 villages in the districts of Sarqalla, Tilako and Awaspi benefited from this programme. 250 of these women were Anfal survivors. Since 2003, however, many NGOs have withdrawn from Kurdistan and have stopped their work in the region as the exceptional circumstances have ended and there is government in place. This has highlighted the shortcomings of the Kurdistan Regional Government in meeting the needs of survivors. Meanwhile women centred developments continue to take place in the region due to the pressure from independent press, tireless campaigning by women's organisations, and more recently because of the competing list for the elections.[24]

---

21   Nalia, June 2010.

22   Acorn's brick making factory project in Garmian failed because it was really difficult work for the women and there was no market for the goods.

23   Ikrama Ghaib, Kalar, May 2009.

24   The Change List 'Gorran' lead by Nawshirwan Mistafa competed with the KDP and PUK coalition (the Kurdistan List) and managed to secure 25 seats in the 2009 elections (out of 110). The pre-election campaign and Gorran's success put pressure on the Kurdistan

Religious affiliation can also offer protection against stress because it can provide 'a form of ideology with respect to which psychological coping mechanisms may be structured' (Ager 1993: 15). In this study religion played the role of a mediator of violent loss and injustice. This was particularly true for older women who had lost their husbands and sons. Many women believed that 'It is just God who helps you survive'[25] and that only God could 'return our justice': heqman bisene. This kind of faith helps sustain the survivor particularly when many perpetrators and collaborators have escaped justice. There is a need to believe that somewhere, in some other world, the people who have ruined one's life and destroyed all that was valuable and precious, would be punished in some way.

Many times when a woman cried abundantly, she would wipe away her tears and remind herself that all that is happened is in accordance with God's wish and that God knows best. God does not want them to cry for the dead, he wants them to accept their fate and live honourably so this is what they try to do.

Ghazala who lost her husband and two sons broke down a number of times while talking about them. Each time she wiped her eyes and repeated, 'Alhamdu Li-llah, Thank God.' To cry like this, she believed, meant being ungrateful to God who had taken away her husband and two children but let the other five survive. A part of her gratefulness seems to be because she is scared of God:

> If we do not thank God, we can't manage. God has given us eyes, legs, hands and he can take all of them away. He took our sons, husbands, cousins ... Every time I feel down I say, thank you God, this was your wish, I am indebted to you. It was God's will that they died this way.

The fear that God has the capacity to do worse things makes some of the older women repent when they cried and felt grieved. Believing in God and destiny may also make some survivors feel less guilty because it makes them realise that they could not have protected their loved ones, they could not have died in their place. They could not have changed the course of events by resisting, fighting, taking risks because 'when God writes something as your fate, no one can change it.'[26]

Women's Empowerment

In many ways Anfal was a process of disempowerment – enduring loss and violence during Anfal and marginalisation and deprivation in the aftermath. Lack of power is of great importance when we are talking about the women survivors of Anfal. Women in this group already experience powerlessness within their families and

---

Regional Government to provide betters services to the community at large but especially to the Anfal affected regions.

25   Shadan, March 2006.

26   Srwa, December 2005.

intimate relationships and losing control over the course of events is a further source of disempowerment.

Dealing with the consequences of Anfal, however, required active and adaptive coping strategies. During my numerous encounters with Anfal surviving women it became clear to me that they are not passive victims. They are active and strong survivors. This is fundamental for anyone who wants to work with or on behalf of these women. This understanding needs to be integrated into theory, policy, and practice. The women have found ways to reclaim power within their communities and rebuild their lives. The exercise of power through agency and freedom in pursuing one's goals are central to the human existence. Burstow (1992, chapter 1) argues that:

> Our human existence is predicted on our ability to project meaning, to embark
> on projects, to create world. As subjects we are forever creating and re-creating
> the world by making new choices and by ordering what is around us.

Similarly Anfal surviving women have reclaimed power by taking advantage of their new found positions, by creating social groups, and by narrating their stories to each other and to the rest of the community. Through narrating their stories the women survivors have voiced their anger and disagreement with the dominant discourse on Anfal. They have drawn attention to their current problems and they have requested help (I will address this in the next section). Over time the women have become heads of families and despite limitations imposed by the patriarchal society, they have been successful workers, mothers, and citizens (see Chapter 5). They also created informal and formal networks to help, support and protect each other and I will now describe these in more detail.

Anfal surviving women found a way to survive their contradictory situation in the male dominated society which, on the one hand, did not provide them with financial security traditionally given to women and, on the other, blamed them when they went out and worked to support their children. They realised that they had to stick together and to defend each other's honour against rumours and stigmatisation. Evidence gathered by other researchers also suggests that the creation of groups is empowering for displaced women (Light 1992, Roe 1992). By forming groups women can organise themselves according to their needs and priorities; they also offer a supportive environment and the possibility of challenging oppression together. Through uprooting, death, and trauma Anfal destroyed the necessary bonds for survival. The importance of forming groups is, therefore, an attempt at repairing the severed connections.

Women's groups can function as a support network or take on more political aims. In the first case, when women get together in a safe space, they typically bond and help each other by sharing experiences and by crying and laughing together. The literature (DeChant, 1996, Harris, 1998) suggests that these processes are central to effective group work with women, and that they are more easily created when the members are disadvantaged than when they are privileged (Williams

2004). Anfal survivors have formed various informal networks to support each other. During Anfal they formed groups to protect themselves from sexual abuse (see section on sexual abuse). In the aftermath they often sought work in groups to defend themselves against possible abuse and rumours (see Chapter 5).

Many women who did not have immediate male relatives that could take care of them chose to live together or close to each other. This is particularly true in the case of women who were related to each other by blood or marriage. Habiba, Shirin and Shara are related by marriage. All three of them are widows of various political turmoil in Kurdistan. Habiba's husband, Shirin's brother, is the only of the three men who disappeared during Anfal. The three women and their children now live in one house in Aliawa village. They share the caring for their children and the income they acquire through each of their work. However Shirin, who is the sister of the other two's husband's, plays the role of a guardian in the family. She makes decisions for the whole group and she is consulted about things. In some ways this reflects the traditional power structure because the other two women who have lost their own husbands are putting their trust in their husbands' sister. Shirin is more authoritative and during the interview it became apparent that she is clever and in control.

A few weeks after deportation the inhabitants of Bahirka were allowed to go out and work. Every morning Ghazala joined women from Chelke and other villages and went to the square in Erbil. At dawn the place was buzzing with men and women, who stood on different sides, waiting to be employed. The employers then chose the workers and provided transport to the fields. The villagers picked whatever was in season, put them in boxes, and carried the boxes to the storage place a few hundred yards away. Sometimes, there was no work, reported Ghazala, because employers only chose the strongest and most experienced workers. Ghazala herself, and other women from her village, had never done this kind of work before and they found it difficult. The women of Geeze, on the other hand, had a reputation for being good workers. They usually got work while other women did not. The Geeze women then asked the women of Chelke to join them. They promised to teach them how to do the job. This helped many inexperienced women obtain work. They were able to feed their children because of this support.

Other groups may have more political aims and aspire to bring about change for women of the community. Jini nwe bo jinani Anfal – New Life for Anfal Women is an organisation which was established in July 2003, after the fall of the Baath regime, by a group of women who themselves were either survivors of Anfal or whose immediate families had been victimised by Anfal.[27] The organisation is based in Kirkuk and aims to raise awareness amongst the women survivors about legal, social and health issues. In 2005 a series of workshops were provided for women survivors to raise awareness about the concept of democracy and the Iraqi constitution. The organisation also aims to raise awareness about the role and status of Anfal surviving women in the community at large in the hope of more

---

27 Lana, March 2006.

recognition and acknowledgement. It also defends the women survivors and aims to get them compensation and financial support.

Some of the women who started this organisation are themselves grappling with issues of recognition and acknowledgement. One of the founding members of the group is Lana. She is an Anfal survivor who lost her husband and baby son to Anfal. Later her in-laws took her children away from her. Since then Lana has worked in various places to sustain herself and has fought to have access to her children. She has bad health and has suffered from severe depression over the years. Now Lana seems to have found solace in working for this organisation. 'I really like my job,' she told me, 'because it is to the advantage, awareness and serving Anfal women in particular.'

Another Anfal surviving women's organisation which was also established in Garmian is dealing with similar problems. A delegation of women came to Berlin for a one day seminar on Anfal on 17 April 2008. Shazada Hussein represented women survivors in the workshop by talking about her experiences of losing her husband, incarceration during Anfal, rebuilding life afterwards, and raising her daughter alone. Shazadah was already a member of Kurdistan Women's Union and she was an active defender of Anfal surviving women's rights. As a result of this trip and the continued conversations that started in Berlin a group of 100 women in Garmian started thinking about a monument to represent the experiences of the women survivors of Anfal (Mlodoch 2009). This involved a lengthy process of women discussing what they want, how they want to be represented and remembered, and then negotiations with the Kurdish government to get funding and support for the project. This is part of the healing process. Starting with women's unhappiness about how the government commemorates Anfal and represents the victims and survivors, the women discussed how they want to be perceived and ended up with reclaiming their own image and status. This is an important step in the direction of empowerment, seeking justice, and closure.

The above examples illustrate a potential route for the empowerment of women survivors of Anfal, women collectively taking charge, helping each other and reclaiming their own voice. These routes allow women to build their inner strength as part of the process of working together for the reconstruction of their lives and communities.

Women's Narratives, Representation and Voice

In the post-1991 era in Kurdistan Anfal survivors became the vehicle of telling the Anfal story. They were interviewed by various local and international TV and satellite channels, government officials, NGOs, and researchers. The resulting Anfal narrative, however, was not in accordance with their experience and expectations. Some women felt exploited by all the different groups and they refused to speak to anyone. They preferred to just get on with the business of daily life because each time they talked they felt that their 'wounds are picked' – 'brinman ekoletewe.'

The majority of the women I met, however, wanted to talk and to tell their story, sometimes in the absence of visible emotions. I believe the fact that I was an independent woman researcher (not connected to the government, media or NGOs), that I had gone back from Europe to listen to their stories, and that I listened well made all the difference. These women have become highly sensitised to being exploited. They complain about how their stories have been used for political purposes and for propaganda. Through telling their stories, feeling well heard, understood, and connected the women created a better understanding of their own situation and remembered their own strengths and resilience. These conversations, despite their limitations, can affirm their lives and experiences and reclaim their stories.

A researcher's sensitivity to the effects of gender, trauma, and poverty on people's lives may help the witness explore other possible meanings of situations by revealing the social construction of the community's expectations and perceptions of them. Narrative helps women shed light on the positive stories of survival and agency that have been overshadowed by internalised negative perceptions of themselves as powerless and hopeless victims. 'Don't you know what women are like?' Semeera told me echoing the society's perception of herself, 'We are weak.' Women may feel empowered by talking about their lives. They recover incidents of creation and strength that might be forgotten in the light of other negative stories, which they have come to believe about themselves.

Anfal surviving women have faced great obstacles to building normal and happy lives. Some women felt hopeless and depressed because of these experiences but talking about their lives helped them begin to realise that these are not personality traits but a natural reaction to extreme and abnormal situations. It is important for women to recognise that these negative experiences are only part of their lives. Talking about times when they have been strong and active reminds them that 'they are strong, they have survived' (Herbst 1992, 148). Kowalenko (1988) states:

> Language externalises as nothing else can … it represents human experience as it has been or is being lived … data from the individual's life are retrieved and pressures are relieved thereby increasing self control. (Kowalenko 1988, cited in Herbst 1992: 148)

Swan (1998) worked with Aboriginal women who belong to a highly stigmatised group. She points to the importance of narrative for reclaiming their voice and fighting negative stereotypes of their people. The author points to stories of strength and survival which have been marginalised and overshadowed by the dominant group. This gave the women an opportunity to break free from destructive stories and enabled them to see themselves differently and take pride in who they were. In this way, narrative has the potential to help the person understand her situation, come to terms with it, draw on her inner strength and history in order to construct a more positive image of herself, and hence enhance self esteem and empowerment.

Women's Anfal Narrative

Women are voiceless, their stories are marginalised, and their experiences considered unimportant (Ringelhei 1997, Leydesdorff et al. 1996, Butalia 1997). Marginalization in turn puts individuals at risk for mental illness (Myers et al. 2005). Here I will talk about women's narratives of Anfal and how these narratives may be influenced and restricted by social and political factors. The three factors that I will discuss here are: The gender expectations of how men and women should feel and behave; the official Anfal narrative; and social stigma around certain experiences.

Gender Expectations

While conducting my fieldwork in Kurdistan I visited villages in the six geographical areas targeted by Anfal. Despite explaining that I was an independent researcher interested in talking to the women survivors various men came forward to talk to me about their experiences. In some cases I had no choice but to interview the men, to get them out of the way, before I was able to talk to the women. Soon it became apparent that there were differences between the testimonies provided by men and women.

Generally speaking the men's testimonies were formal and impersonal. They were good in providing information about the chronology of events, the locations where the attacks were launched from, the dates of attacks and names of places. On occasion if a man became emotional he would stop talking until he regained control. Older men were more likely to cry as they remembered and talked about Anfal. In one case a man who did not cry for his baby son's slow and gruesome death as a result of a gas attack, cried when he started talking about his comrades. He cried for the loss of the revolution and 'all the good peshmarga who were martyred.'[28] He argued that he cried for them because these men died for Kurdistan, it was a choice they made. 'These men were patriots,' he said, 'they were my friends in struggle, they were brave men.' I could not help but think, however, that he needed to provide a more socially acceptable reason to cry because weeping for his own son was not acceptable.

Rizgar[29] lost six brothers, two sisters-in-law, six nephews and nieces, one sister, his mother, wife and two children. He talked in detail about various issues around Anfal, sometimes providing lengthy details about marginal issues. It was very difficult, however, to get him to talk about his own personal loss in the process. At one point when he became tearful he stopped talking until he regained his balance. It was obvious that he found it difficult to connect with his emotions about the extreme loss he had endured and he dealt with this by keeping continuously busy.

28 Hameed, June 2010.
29 Rizgar, March 2010.

He worked as a member of the village council and travelled to various locations every day, working very long hours.

An elderly couple talked about losing their three sons.[30] The woman could not tell the story because she was crying. Her husband, who remained calm and seemed detached, kept telling her, 'Stop howling, no amount of sobbing will bring them back.' Socially, women are expected to be emotional while men are expected to be in control. This in turn affects what men and women will recall and how they behave while recalling their memories.

The women provided more personal and emotional testimonies. They expressed rage, regret, and sadness. Many of them cried while remembering. Women are expected to care about others, to put others' needs first and to identify with them (Rampage 1991). In this research women were more willing to speak about the suffering of others – men, women and children whom they had met during their journey and incarceration, and to cry for them. This is probably best explained by Tonia Rotkopf-Blair, a Holocaust survivor, who states that even though she comes from a liberal family she has internalised the public views about a woman's place in society and this affects what she remembers about her own suffering: 'It is interesting that one doesn't remember one's own problems so much: hunger, etc. But I do remember the others' hunger and pain much better' (cited in Tec 2003, 127).

Unlike survivors of other cases of genocide and mass violence, in particular the Holocaust, most of the women survivors of Anfal are illiterate or semi literate. They, therefore, cannot write their own memoirs or make their own point of view public. The only way they do this is through providing testimonies, by talking to each other and to various strangers who pass through collecting information for various purposes. However, it is men's accounts that are privileged and they are the ones who shape the dominant Anfal narrative.

The Official Anfal Narrative

Another factor that influences memory is the political discourse on Anfal. In the dominant narrative Anfal is defined by death and destruction. It concentrates on people who were shot in the mass graves, the destruction of 4,000 Kurdish villages, the extensive use of chemical weapons, the crushed Kurdish liberation movement, and the fact that genocide was carried out against the Kurds. For a long time the Anfal commemorations involved interviewing survivors who seemed to be crying forever and talking about their disappeared loved ones. It also involved showing images of the devastation (gas attacks, gas victims, leveled villages, and mass graves), televising the yearly ceremonies, and broadcasting politicians' speeches. More recently it included showing images of the (forever) falling statues of Saddam Hussein, images from his capture, footage showing Ali Hassan Majeed inflicting pain on and shooting people, and images from the Iraqi Special Tribunal.

---

30   Hajar, April 2006.

This focus on death, disappearance and violence is similar to other parts of the world where mass violence has taken place. When talking about the South African Commission Fiona Ross (2001: 252) states:

> The definitions of violation set out in the Act are largely to do with what can be done to the body – it can be abducted, tortured, killed, disappeared. In the commission's hearings the main focus was on bodies and on the visible embodiment of suffering ... this has important implications for the ways in which women's testimonies can be heard.

Similarly a large part of women's experiences during Anfal are considered irrelevant. Even the focus on bodies is usually a focus on men's bodies. Women's battle with hunger, filth, menstruation, illness, birth, death, and sexual abuse in the prison camps, the grimness of life in refugee camps and in hiding while internally displaced, and the traumas and illnesses associated with gas exposure are not given space in the Anfal narrative. These are seen as insignificant compared with men's more extreme experiences of torture and death. In a study about the women survivors of the atomic bomb Todeschini (2001: 104) addresses the politics of expression and argues that:

> The public acknowledgement and appropriation of illness, loss, pain, and grief, and the establishment of gender categories in relation to suffering, are profoundly political acts, which draw boundaries and determine 'appropriate' expressions of suffering.

In the dominant Anfal discourse the 'appropriate' experiences and remembrances of Anfal do not include gendered experiences of violence and suffering. This influences what women would or would not talk about. In this study women were generally silent about some of their intimate experiences. No one spoke about problems associated with women's bodies spontaneously. Even when I asked them questions they usually gave brief answers.

Along the same line the 'appropriate' remembrance is to talk about what happened during Anfal and not after. The narrative centres on what happened between 23 February and 6 September. It does not concern itself with the consequences and aftermath which are widely spoken about by women survivors. This focus on the immediate violence of a catastrophe seems to be common in other parts of the world. When talking about women survivors of the atomic bomb Todeschini (2001: 105) states that the official discourses are often in conflict with women's interpretations:

> In contrast to the more conventional testimonial accounts, the main thrust of [the women's] narratives is not so much a preoccupation with the atomic explosion itself and its immediate aftermath but with what came later, the years and decades after the bomb.

Anfal surviving women talked at length about the aftermath of Anfal in their testimonies. They talked about their preoccupation with survival and their struggle while single-handedly raising children and making ends meet. Many of them complained that when visited by journalists and government officers they regularly tried to draw attention to their current problems but most of the time this part of their story was edited out in the documentary. Some survivors are fed up with being interviewed. They need practical support in their lives. They don't want to keep talking about the awful things of the past. Behe talked about Kurdish TV channels: 'They keep our pain alive. Every time they show Anfal we live the experience again, our back breaks.'

Women talked about Anfal being used as a weapon by the Kurdish politicians as a means of encouraging people to vote for the Kurdish government in the Iraqi elections. The TV channels were broadcasting heartbreaking interviews with survivors and images of gassed women and children. This was to remind people what they endured under a non-Kurdish government and to encourage them not to repeat the same mistake. The message inherent in these images and interviews was clear: Vote appropriately, don't let Anfal happen again. Kurdish politicians also used Anfal in their negotiations with the Iraqi government and their appeals for help and sympathy to the world without providing effective and crucial support to those who were victimised.

Social Stigma

Women were reluctant to present some of their intimate experiences because of the social stigma that may arise if they spoke. They did not feel entitled to complain about problems associated with their bodies. Traditionally even women's undergarments are not allowed public exposure. When putting washed clothes on the line my own mother used to hide her undergarments under other clothes. It was shameful to hang these clothes in public. Women's bodies are taboo. They are mysterious and shameful objects that need to be covered up and protected.[31] Talking about these bodies in public, what they endure and what can be done to them, is not acceptable.

In many communities women are held responsible for the crimes committed against them (Das 1997). The majority of the Anfal women I interviewed were not prepared to address sexual violence (see Chapter 2, section on Sexual Abuse). The silence about sexual abuse is, however, gradually being broken. Women's organisations, for example, have begun to address the issue of sexual abuse as did the Public Prosecutor in the Anfal trial who spoke about widespread rape and abuse.[32] As these issues make their way into the dominant discourse and become

---

31   Kurdish folklore songs are full of erotic descriptions of women's bodies. This only contributes to the perception of women's bodies as sex objects. As a result women find it embarrassing to talk about their bodies.

32   International Centre for Transitional Justice (2007a).

'acceptable forms of suffering' women become encouraged to talk about them. Even now, it is easier for older women to address these issues or activist survivors who take on the battle on behalf of others. The majority of the women remain silenced by traditional social concepts of shame, honour and dignity.

Addressing Women's Needs

Dealing with survivors' needs in the aftermath of a national catastrophe is no easy thing. This is especially difficult when a major shift in the political climate takes place. From 1992, when the Kurdistan Regional Government was established, the Iraqi government lost control over a major section of Kurdistan, in particular the provinces of Duhok, Erbil and Suleimanya and the surrounding districts. A large section of the Garmian region was also under Kurdish control. The city of Kirkuk, however, and some of the towns and villages around it remained under Iraqi control. The Iraqi state had no interest and no intention in supporting the survivors of the Anfal campaign. This left the newly established Kurdish government, with all its problems – lack of experience, shortage of fund and administration – to deal with the aftermath of the catastrophe.

In the early 1990s, due to the international sanctions on Iraq and the Iraqi sanctions on the Kurdistan region, poverty was widespread in all sections of society. Sometimes the Kurdish government was unable to pay salaries on monthly basis and some families had to learn to survive on one month's salary for three months. This was at a time when inflation was high and unemployment was widespread. The Kurdish civil war, between 1994 and 1998, further complicated the situation when other, more urgent issues concerning current violence and conflict continued to sideline the rightful claims of the Anfal survivors for compensation and support.

Over the years things gradually improved due to UN's Food for Oil programme in 1997 and the Kurdish government's attempts to support the survivors by providing a minimal salary and some housing (see Chapter 5). These responses from the Kurdish government were in many ways too little too late. As the children of the survivors have grown up and some of their basic needs have been met their complaints have gradually shifted from urgent requests for help to quest for recognition, closure, and justice. Yael Danieli (2007) points out that while making reparations to victims and survivors, three things need to be taken into account: Financial compensation, bringing perpetrators to justice, and community acknowledgement and change. It is, therefore, important to have a holistic approach and to tackle these issues mutually.

Most of the women feel a great sense of injustice which, at times, is solely directed at the Kurdistan Regional Government. They are angry with the Kurdish authorities for not giving them enough financial support, for not finding the bodies of their loved ones, for not helping, not listening, not caring, and not reinforcing justice. Turner et al. (2003:198) stress the importance of providing information to survivors about what happened to their families and their communities and for those who are responsible to publicly acknowledge their actions, apologise, and display contrition. This is important because 'if ignored the psychological burdens may simply become entrenched and affect future generations in a cycle of violence and retribution.'

Most of the survivors' demands are directed at the Kurdish government partly because they perceive the Iraqi government as a distant entity which does not directly interfere with their lives. Also because, despite the change of regime and despite the coming to power of more liberal forces in Iraq, which includes strong Kurdish elements, the survivors perceive the central government as hostile. They complain about the Kurdish government because they feel these are the parties they supported and this is the government that came to power as a result of their sacrifices. To them, the Kurdish government is their government and therefore it should be more supportive to them. The role of the Iraqi government, however, cannot be ignored.

The Iraqi government has not started compensating the survivors for the loss of their livelihood, their homes, property, animals and above all their families and communities. On 24 June 2007 when the Iraqi Special Tribunal issued a verdict on Anfal 'the trial chamber noted that complainants had the right to take a compensation case to the civil courts.'[33] The process of compensation for the claimants has not yet started and there has been no recognition that all of those who have lost resources and family members during Anfal, and not just the claimants, should be compensated. It is also essential that the Iraqi state provides an apology and facilitates provision of information about the manner in which their loved ones perished and where they are buried. This, however, became increasingly difficult in the aftermath of the 2003 invasion as destabilising forces were unleashed within Iraqi society and the terrible security situation made the uncovering of the mass graves hugely difficult.

In 2005 the bodies of 500 Barzanis who were massacred in 1983 were uncovered from the south of Iraq and returned to Kurdistan. The search for and recovery of the bodies was captured in a documentary by Gwynne Roberts 'Saddam's Road to Hell.' Many women who had seen the procession of the bodies to Kurdistan questioned why the Kurdish government does not do the same for the Anfal victims: 'We want their bodies to be returned just like Masud Barzani brought the Barzanis back.'[34] Over the last two years, however, more Anfal mass graves have been uncovered and some of the victims have been returned to Kurdistan. Except from the few cases where the victims carried identity cards the majority of the recovered bodies have not been identified thus making it difficult for people to have closure.

In their quest for justice women also talked about the Kurdish collaborators, the mustashars, who took their husbands, sons, and brothers away. Behe recounted 'it

---

33   International Centre for Transitional Justice (2007b).
34   Behe, November 2005.

was the mustashars that did this to us.' In some cases the mustashars persuaded the men to surrender and promised the women that they would return after a maximum of 3 days but they never came back. The women survivors want these mustashars to be put on trail. The problem is not only that in return for their support for the popular Kurdish uprising the jash were given an amnesty in 1991 but the fact that they have become powerful and influential in Kurdish politics. Each mustashar aligned himself with one of the Kurdish parties after the popular uprising and in so doing he secured power and wealth: 'Now they are doing better than us, they are looked after.'[35]

Finally it is important to address the low social status individuals are subjected to by changing the way the official discourse and dominant culture portrays these women. The yearly Anfal commemorations and popular culture have not managed to create a healthy understanding of Anfal and they have not made the society respect the survivors. They have not made the nation think, create, research, and theorise. With some exceptions the majority of the art that is produced is an emotional and non-imaginative reproduction of the violence, portraying civilians as passive victims. Many of the paintings that address Halabja and Anfal are full of soldiers' boots, smoke and skulls. The plays and dramas are about attacks and people running and falling.[36]

It is also necessary to give training to the women and their children and to help them better their lives by providing education and employment opportunities. The fact that many children were deprived of education meant they ended up taking the lowest paid jobs which is now being held against them. They are perceived as uneducated and backward individuals whom society wants to break away from (see Chapter 5). In the rush towards liberalisation and progress the survivors and their children are sometimes seen as reminders of a violent past which society wants to leave behind (Mlodoch 2008). It is therefore important to raise public awareness about survivors' needs and also to help them get involved in the modern social and political environment. So far the survivors are not represented in senior public sector roles, the parliament, and the government. It is important that, at least, the children of the survivors are helped to achieve their potential in order to better social cohesion and the reintegration of survivors into Kurdish society.

Conclusion

Most individuals exposed to extreme events and conditions are remarkably resilient (Solomon 2003: 7). The majority of Anfal survivors I talked to were able

---

35   Behe, November 2005.

36   Some brilliant exceptions are Sherko Bekes's book length poem Butterfly Valley, Bachtiar Ali's novel The City of White Musicians and Omsan Ahmad's artwork (some of which were in the exhibition at the Imperial War Museum, Displaced).

to adapt and recover once peace was restored and they had rebuilt their lives and communities. Others suffered from ongoing psychological reactions to Anfal and a smaller group seemed to have disabling mental health problems. For this group it is important to provide community development programmes alongside specific therapeutic programs. The immediate reactions to Anfal included fear, flashbacks, numbness, helplessness. Later, as years passed, survivors seemed to experience a different range of psychological reactions such as depression, insomnia, rage, memory problems, muscular pain and quest for justice, compensation, and closure.

Women in this study used various strategies to deal with Anfal and its aftermath and some fared better than others. Some of them seemed to be fixated on the past and they could not escape their memories or stop comparing their current situation with life before Anfal. Some reported going to funerals in order to process their grief in public. Others seemed to prefer to get on with their daily lives and not talk or remember. Some women refuse to give interviews because talking about the past exposes them to remembering and sometimes re-experiencing it. Then they are left alone to deal with the aftermath of these recollections (Levine 1995: 84, Matussek et al. 1975: 47). The majority however, chose to speak out, to defend themselves, to express rage and to demand help. All the women survivors have coped with extraordinary circumstances and have managed to look after themselves and their children sometimes at great cost to their health and happiness.

Support provided by family, relatives and the community was experienced positively. The same goes for having faith and believing that they should accept God's choice even if they don't understand it because God will eventually give them their justice. Turning to other women who are struggling with similar issues was also beneficial. Women provided informal support to each other and, as part of their developed group identities, they felt responsible for each other. Other group formations also helped through reclaiming their voices, demanding their rights and fighting for change.

Women's testimonies were different from men's in terms of the personal information they gave of themselves and others and the emotional quality of their remembrances. The social gender expectations influence women's testimonies as do the dominant narrative and social stigma. Women were more likely to express disapproval about the way their story was represented when it came to curbing their complaints about the Kurdistan Regional Government (seeking financial support, housing and the attainment of justice – Kurdish perpetrators). They also expressed anger for being used by the Kurdish government in negotiations with Iraq and during the elections. They demand that problems associated with their current socio-economic conditions should not be ignored. Like others in the community most women were unwilling to talk about their experience of sexual abuse and generally problems associated with women's bodies. This, however, is changing as more official channels such as independent media, women activists, researchers, and the judiciary is starting to talk about these issues. Once sexual abuse becomes spoken about in the public realm it becomes possible for individual women to begin talking about their personal experiences.

Chapter 7
Learning from Women Survivors of Anfal

The core experiences of psychological trauma are disempowerment and disconnection from others. Recovery, therefore, is based upon the empowerment of survivors and the creation of new connections. (Herman 1997: 133)

There is no justification for thinking of the women survivors as simply victims of Anfal. Detailed understanding of women's experiences enables us to appreciate their efforts to survive individually and collectively. This book has explored the traumas women experienced as a result of Anfal through gas exposure; uprooting; witnessing violence and destruction; loss of possessions, family, community, and means of employment; and dealing with poverty, stigma, and exclusion in the aftermath. It addressed the experiences of the various groups of women whose lives were forever changed by the campaign. The findings will now be summarised followed by a discussion of the women survivors' image in the media, the disconnection between the survivors and the rest of the community, their pursuit of justice, and obstacles to mourning and closure. These will form the basis for making recommendations for change.

Providing the historical context to the Kurdish situation in Iraq, Chapter 1 went on to outline the eight Anfal attacks targeting six rural regions where the Kurdish resistance was most active. The Iraqi state's motive behind the September 1988 General Amnesty was studied followed by a discussion about whether or not Anfal was genocide. After examining the coverage of the campaign in 1988, the reasons for the international community's silence and lack of acknowledgement were looked at. The chapter concluded by arguing that Anfal was genocide and that silence around it was due to converging international interests.

Chapter 2 addressed the issues faced by women who were arrested and detained in the camps. Disorientation in the face of bombardment and gassing, forced displacement, separation from family members, and the conditions of life in detention were recorded here. This was followed by women's more intimate experiences of sexual abuse, giving birth, witnessing death, and receiving help. It was concluded that despite the imposed powerlessness camp life did not amount to passivity. Some women reported cooperation through sharing money, looking after the ill, and protecting each other from possible sexual abuse.

The situation of refugees and IDPs was looked at in Chapter 3. The refugees' journeys were described followed by life in the refugee camps where people lived under continued threat of danger as a result of the bad security situation and the fear of Iraqi attacks. IDPs, who had gone into hiding during the campaign, were described as more vulnerable because of their complete dependency on their hosts, the continued fear of being betrayed by their peers, and because of being deprived of the support that refugees were entitled to. It was found that some risk taking behaviour (attempting to flee against all odds, for example), luck, and support by others were important for surviving. Both refugees and IDPs, it was concluded, experienced lack of control and both groups tried to cope with their situations by moving to find more liveable and secure environments.

In Chapter 4 the ordeal of the gas survivors was addressed. It was found that chronically ill women were stigmatised in the community as sources of 'contamination.' Descriptions of women's experiences during the gas attacks was followed by outlining issues around ill health, beliefs about their health problems, social stigma, psychological problems, and social consequences. It was argued that since our knowledge about the long term effects of mustard gas and nerve agents is incomplete, survivors experience intense fear and stress in the face of the unpredictable health consequences of these weapons. In the absence of community, government, and international support women continued to suffer and some of them died of persistent health problems years after exposure.

Chapter 5 presented the challenges of rebuilding life after the devastation and loss of male breadwinners. Women, who were overburdened with the responsibility of caring and providing for their children and elderly dependants, did the most labour intensive and lowest paid jobs. They were exploited in the community by various groups and individuals. Poverty, stigma, and their inability to support their children's schooling made women feel guilty. The role of women's organisations and the Kurdish government in helping these women was also discussed. It is concluded that the lack of a holistic approach to tackling Anfal surviving women's needs has caused great pain.

The mental health consequences of Anfal were discussed in Chapter 6. The range of mental health problems that survivors experienced included fear, traumatic pain, numbness and flashbacks, grief, depression, isolation, contaminated memory, embodied distress, anger and the need for resolution. It was found that a number of factors such as social support and religious affiliation were found to be protective against trauma. Similarly, women's empowerment through creating formal and informal groups, and speaking out to reclaim their stories were looked at followed by women's demands for compensation, closure, and justice.

Representation in the Media

Despite the horror, loss and trauma of their lives, I was endlessly reminded of women's agency, and their effort to survive. In contrast, our dominant cultural imagery – promulgated through the media – is of women as victims. Dominant images for me are those of gassed women and children whose faces are blistered and blackened, Anfal widows beating their faces and raising fingers to show the number of relatives lost, weeping and desperate women talking about Anfal, a lamenting old woman in an empty house. The widespread presentation of the mutilated bodies of the gassed civilians (from Halabja and the Anfalled villages) and images of Anfal widows who are dressed top to toe in black and seem to be frozen in grief have had negative consequences for the survivors and the community at large.

I suggest that there are three important implications of these types of broadcasts. First, the repetitive transmission of these images is counterproductive. If the purpose of showing them is to stir people's emotions and to make sure that Anfal is not being forgotten, these images have gradually lost meaning and their abundant presentation has led to compassion fatigue. They have lost their value in terms of national commemoration. Many people in the community are fed up with these images and they change the channel whenever they are broadcast. The images are not informative of anything new and they have lost their psychological impact and historical significance in the consciousness of ordinary civilians.

Secondly, representing women as grieving widows who are always lamenting for their loved ones makes them look like eternal and powerless victims. In this sense those who were victims of a nationalist central government are once again being victimised within their own community. This kind of representation of Anfal women disempowers them and reduces their value to that of mere victims. This is neither fair nor healthy. It is not fair because it oppresses survivors and disvalues their abilities. Contrary to this representation, it can be argued that Anfal surviving women changed the definition of 'woman' in the Kurdish society. If women were perceived to be powerless and feeble individuals who needed a man by their side to get on in the world, a man who would fend for them and protect them from social dishonor, the women survivors of Anfal proved that they can survive without men, that they are able, independent, intelligent, and powerful. This kind of representation is also unhealthy because it may influence survivors' own perceptions of themselves. They too may learn to see themselves as mere victims. They may feel sorry for themselves and experience lack of confidence and low social status. This prevents them from seeing their own strength and achievements.

The third consequence of showing these images is that, like all other horrific images of mutilation and violence, they enter the social psyche and become normalised. Such violent images that become part of daily life shape the imagination and may even be reproduced in various forms. This, in turn, may promote tolerance of violence in society. The rising number of 'honour' killing, violence against women and children in Iraqi Kurdistan in the post-liberation era may be one of the consequences of the past political violence and its vivid reproduction in common culture.[1] A more responsible approach is required to deal with Anfal in the media which will not victimise the survivors, traumatise the community, or contribute to the reproduction of violence.

---

1   More research is required in this field.

The Severed Connections

Encounter with extreme violence shatters the individual's faith in a fair and predictable world. It isolates the individual from her family, tribe, and community. Herman (1997: 51) explains that traumatic events 'have primary effects not only on the psychological structures of the self but also on the systems of attachment and meaning that link individual and community.' On top of the damage caused by the violence the women survived many let-downs and betrayals which further isolated them. These betrayals were more strongly felt when members of one's own community were involved. The role of the jash in collaborating with the government, bringing the army to the remote hideouts, announcing false amnesties, and promising protection to civilians is reported with great anger and resentment. The anger is particularly fuelled by the fact the jash were granted amnesty by the Kurdish revolutionaries in 1991 and some of them have become powerful players in current Kurdish politics.

The inability of the Kurdish revolution to protect civilians during Anfal caused disappointment and anger. This was particularly true in cases where regional commanders prevented people from fleeing when there was still an opportunity to escape, hoping that civilian presence will save their lives. The anger and disappointment were enhanced by dissatisfaction with the Kurdish authorities in the aftermath of Anfal because they failed to support the village population that had supported them throughout the revolution. On 16 March 2006 when riots broke out against the Kurdish authorities' lack of concern for Halabja survivors one of the raised slogans said: 'Eme bo ewe mirdin, ewe chitan bo kirdin? We died for you, what did you do for us?'[2] This is an example of the widespread resentment felt by survivors who believe that they are neglected by the same people that came to power as a result of their sacrifices.

Because women's lives are primarily lived through connections, the betrayals and let-downs by friends, relatives, and acquaintances, who failed to help these women at moments of need, were reported with great sadness. This contributed to the erosion of trust and connectedness and more experienced isolation. To this must be added exploitation by employers, landlords, and other individuals in the aftermath which are less openly reported by individual survivors but are commonly known to have taken place in the community. The stigma around women as heads of families, around 'the fatherless Anfal children' as well as poverty and voicelessness are major sources of grief.

Women found it difficult to trust after Anfal. Initially, in the aftermath of the 1991 uprising, they started telling their stories to the public. Some of them talked to create an understanding of their situation and in the hope of attracting some support. Some talked as a means of processing their experiences, because it was an outlet for their grief and it was experienced as therapeutic. Others talked out

---

2   I was in Halabja on this day and for the first time I realised how serious people's anger and disappointment with the Kurdish government is.

of a sense of national consciousness, being aware that what they had endured during Anfal is part of the nation's history and it should be recorded and not forgotten. Soon, however, they were disillusioned about receiving help, creating an understanding, being included in the Anfal narrative, and feeling relieved. The testimonies now are more about their dissatisfaction with the status quo and their anger because of lack of redress, support, sympathy, and justice.

Pursuit of Justice

Anfal was all about injustice. It dominated and isolated individuals. It destroyed, deprived, caused injury, grief, stigma, and poverty. Herman (1997: 178) states that at some point all survivors reach a point where 'all questions are reduced to one, spoken more in bewilderment than in outrage: Why? The answer is beyond human understanding.' What survivors experienced during Anfal was appalling, unjust, and unexplainable. Helplessness and outrage were normal responses to failing to understand why they were subjected to such atrocities. As time passed, however, survivors established safety, regained control of their shattered lives, and found new support networks for themselves. They then became more concerned with justice.

Justice has several dimensions for survivors. Addressing marginalisation and inequality is a major element in their quest for justice. This is particularly true in the case of women seeking equality, acknowledgement, and inclusion for their children in the Kurdish society. They demand that their children's low socio-economic status, limited level of education, and their current positions as poorly paid workers should be seen as a consequence of the Anfal campaign and it should be addressed by the Kurdish community and government.

They also seek compensation and support for their losses and grievances. They address their complaints and requests to the Kurdish government, because they believe this government is their own and it should respond to their demands. They, however, fail to recognise that 'compensation' should be provided by the central Iraqi government and what they receive from the Kurdish government is merely social benefit to support the survivors. It is the KRG's responsibility to put pressure on the Iraqi government to compensate the survivors. In the context of continuing conflict in Iraq and the increasing number of victims this may prove to be difficult. Nevertheless, it is important that the government is perceived to be attempting to defend and secure their rights.

Another major request is the improvement of living conditions in the villages and housing complexes where survivors are living. The provision of basic services such as paved roads, clinics, schools, water, electricity, libraries and entertainment facilities are crucial to address the general inequality experienced by survivors. The equitable distribution of resources is essential to tackle social deprivation. Blas et al. (2008: 1685) point out that reducing poverty and reversing exclusionary processes enhance social cohesion and better population health. This is also confirmed by Belle and Doucet (2003: 105) who state that 'it is the most egalitarian societies, not the wealthiest societies that have the longest-lived citizens.'

Finally, survivors demand that the jash leaders who facilitated the disappearance of their men should be brought to justice. They want them to be tried in Kurdistan where the survivors can go and give testimony. Addressing this issue is essential if survivors are to feel that they are valued members of the community. It is also significant because, so far, these men have been granted unconditional immunity and support by the government. Most of them are wealthy and untouchable. This, ironically, makes survivors feel that they are being punished for their support for the Kurdish government and the jash leaders are being commended for their betrayal. These feelings should be taken seriously and dealt with before they lead to further alienation and eventually violence.

Closure and Mourning

Survivors had no time to mourn their losses after Anfal. At first the sudden social change meant that women had to engage in the daily tasks of survival, making new homes and working long, exhausting hours which left them with no time and energy to mourn. They also harboured hope about the return of their loved ones. Women reported having hope until 2003. Only after the fall of the Ba'ath government and the release of political prisoners they became certain that their loved ones were not alive anymore. Still, not having a grave to mourn meant there was no closure.

Herman (1997) points out that the process of mourning can be so painful that survivors try to avoid it. The author points out that resistance to mourn can take on various disguises such as desire for revenge, forgiveness, and compensation. The desire for revenge was reported by many women who wanted Saddam Hussein to be 'hanged' or 'cut to pieces.' Some also wanted revenge on others who were involved in the campaign, wishing that they too experience the same hopelessness and terror as the Anfal victims had experienced all those years ago. But as Herman (1997: 189) points out 'revenge can never change or compensate for the harm that was done.' Similarly, the desire for compensation, though completely justified, may present another impediment to mourning. This is because a prolonged pursuit of compensation may delay the process of mourning and cause more pain and aggravation.

It is essential that survivors manage to mourn their traumatic losses because not mourning means that the trauma remains unresolved. The only way to move forward with grief is to first acknowledge it, accept that nothing can change or compensate for what has been lost, and process the emotions related to this loss. Herman (1997: 190) draws attention to the fact that 'Mourning is the only way to give due honour to loss; there is no adequate compensation.'

Recommendations

There is no hope to move forward if we do not understand the past and deal with its repercussions in the present moment. This is an attempt to understand the meaning and consequences of Saddam Hussein's largest crime against the Kurds in 1988, his biggest violation of human rights in Iraqi Kurdistan. By attempting to reconstruct the past through speaking with survivors we are allowing their understanding to inform and shape national consciousness and, hopefully, national strategy. In this section I will provide a set of recommendations for addressing the current problems for the survivors. These will identity the social changes required to enable women to be empowered through healing relationships, feel that justice has been done, mourn and achieve closure.

Restoring the Public Image of Survivors

Kurdish media must revise its policy towards and perception of survivors. It is essential that different kinds of programmes and documentaries are made which do not concentrate merely on the Anfal story. A survivor's story should be explored from the beginning to the end in order to re-establish continuity and connection in her life, to acknowledge her as a unique and strong individual who has survived against all odds, to highlight her resilient qualities such as generosity, sense of humour, hospitality, resourcefulness, wit, her success at mothering, rebuilding, making a home. Restoring the image of survivors is essential to avoid freezing them in the moment of victimisation and hence further disempowering and victimising them.

It is also important that awards of courage, bravery, and respect are granted to women survivors. These ceremonies should be held in public where the media and members of the community are present. The community must be educated about Anfal and the survivors through well informed and healthy public awareness raising campaigns. It is also important that Anfal becomes part of a wider education curriculum so that future generations can have a wider understanding of the impact of Anfal on their community. This will allow them to analyse such atrocities more critically and objectively.

Another resource that can be taken advantage of is the World Wide Web. Survivors can use the internet as a platform to reclaim their story and image. Grieco (2006) speaks about 're-integrating history' where individuals and communities use the web to communicate their marginalised experiences. These resources, the author points out, can be used to complement, and sometimes counter, the dominant or accepted version of a historical period: 'The patterns and paths of shareable knowledge are no longer confined to the traditional institutions of education ... it is precisely the distributed character of the archive which can be utilised to more accurately record history.'

Giving Back Control and Restoring Faith

It is essential that the women survivors are consulted about their demands. This will give them control by providing an opportunity to express their requests and shape their environment according to their needs. In the aftermath of Anfal, survivors reclaimed power and reconstructed many aspects of their lives. They need further support from the community and government to achieve greater equality, independence, and voice. The media and civil society organisations must play a role in providing this platform for the survivors.

The Kurdish Government must found committees and groups that visit the different regions, listen to the people, and respond to their demands. There may be aggressive confrontations to start with as survivors feel that they have been neglected for too long. However, this has to be faced and dealt with in order to open channels of communication and trust building. This is essential to restore faith in the government. The survivors' basic needs should be incorporated into public policy. They should be given the opportunity to give advice and feedback about government initiatives. Mutually designed programmes should be supported for meeting basic needs.

Attempts at Closure and Redress

Serious attempts should be made to bring back the bodies of the disappeared. If there are obstacles to uncovering the mass graves in Iraq then survivors should be kept informed. Self-help groups should be supported to identify their own communal forms of closure and mourning. The monuments and galleries that arise out of survivor group initiatives are important symbolic gestures. They are more likely to be experienced as therapeutic as compared to monuments imposed by ministries. In the past survivors have rejected these monuments and do not identify with them.

Training and education should be provided for the children of survivors to make sure that they are represented in various sectors of the society as teachers, engineers, doctors, lawyers, politicians, administrators, and civil servants. It is important to secure them a certain percentage of representation in all governmental and non-governmental organisations to combat marginalisation and also to give them the opportunity to defend the rights and needs of their communities.

Mechanisms must be created through which the community can come to terms with its past. Truth and reconciliation commissions may be a start so that the injustices and crimes of the past are not ignored and forgotten. An acknowledgement of these crimes, providing an apology, and information about those who disappeared are essential first steps towards reconciliation and forgiveness.

Social Change: An Aborted Opportunity

A question that remains to be answered is why, despite all the women coming out to work, social change was not achieved in Kurdistan. The death and disappearance of tens of thousands of men during Anfal meant that women had no choice but to break away from traditional gender roles and to participate in economic life for their own survival and that of their children. They entered the job market and took charge of their lives. They built new communities which did not reflect the traditional power structure of the past. Financial independence led to a greater form of independence.

There was an opportunity for women to become decision makers and leaders in their communities, to influence the legal and social systems, and to make progress in the attainment of gender equality. Nevertheless Kurdish women, unlike Western women in the aftermath of the two World Wars, did not manage to change the social order and play a central role in Kurdish society and politics in the aftermath of Anfal. I suggest that there are four main reasons for this lost opportunity. First, Anfal affected a section of the Kurdish community, namely the rural areas. The proportion of the women who entered the job market was small compared to the whole community. In the public sphere, men were still in the dominant majority. They continued the traditional patriarchal system ignoring women's resourcefulness and their role in rebuilding the community.

Second, many of the Anfal surviving women were illiterate or semi literate and they were doing the lowest paid jobs. They comprised the poorest and most marginalised group in the society. They were outside the decision-making and law-making centres. Third, political instability in the region, in particular the Kurdish civil war (1994-1998), destroyed the safety and security required to achieve change. Finally, the strengthening of tribalism in the Kurdish community led to chaos and an increase in violence against women.[3] The community's regression to tribalism had a long history when the Ba'ath regime revived the dying order in the seventies and eighties by giving the tribes money and weapons to fight the Kurdish resistance movement. There were many armed men in the chaotic post-liberation period of 1991 and no law and order. The political and social instability meant that the opportunity for social change was aborted.

Conceptualising the situation of Anfal surviving women must involve an appreciation of the dialectic involved in each moment. Survival confronts victimisation. Creation faces up to destruction. Voice talks down to silence. Reconnectivity bridges dislocation. As women, as survivors of political violence, and as poor people the informants of this research have experienced powerlessness, marginalisation, and stigma. Nevertheless they have fought back by managing the

---

3   Research carried out by various Kurdish women's organisations show that Honour Killing went up in the post-1991 era. This makes sense in the context of free floating armed men in a fluid environment without law and order. In such situations women are easily victimised.

tasks of survival, working in a society which views them with suspicion, creating new homes, social and kin networks, nurturing thousands of children alone in the face of abject poverty, and in the absence of acknowledgement and support, Here, they give testimony and demand justice, through their power to confront the powerful.

Afterword and Personal Reflections

The trouble is that once you see it [the problem], you can't unsee it. And once you've
seen it, keeping quiet, saying nothing, becomes as political an act as speaking out.
There's no innocence. Either way, you're accountable. (Roy 2001: 7)

As women we have been trained from childhood to listen and pay attention to other
people's needs. We have learnt to watch, empathise, and be responsive. Through
this research I came into contact with great suffering. I spoke with numerous
women who had survived violence, abuse, hunger, stigma, and illness having
lost their husbands, brothers and sons. They had lost their property and means of
livelihood and were dumped on the fringe of cities, comprising the poorest and
most neglected group in the community.

I had been living in the UK for 13 years when I started my fieldwork in
Kurdistan. In this time I had completed my education and established a stable life
for myself. I had a job, was gaining recognition as a poet in English, and I lived
a peaceful and content life. At this stage, when everything was nearly perfect and
when I believed the uncertainties and troubles of the past had finally ended, I
launched into this research completely naively. I did not realise that listening to
women talk about what Anfal had done to their lives would change my life. I was
unprepared for how opening myself up to these issues would alter my perception
of the world and of my role in it.

Some of the women I met were severely ill as a result of long term effects of
gas weapons. Others were unwell due to chronic trauma, years of widowhood,
poverty and hard labour. Most of the women were strong and vibrant but the
trauma of Anfal was still fresh for many. At times I felt exhausted and numbed by
the numerous stories of pain and devastation. Sometimes I had to keep reminding
myself that no matter how many times I heard an Anfal story, it was always true
and unique for each individual. Sometimes I found it difficult to believe what I
heard but I now realise that I was trying to protect myself.

We live our lives depending on continuity and routine. We would like to believe
that our world would not be turned upside down, that we would not be tortured
and imprisoned, that we won't outlive our loved ones. We also take some human
values for granted. We want to believe that people are generally rational and good,
that the evil and irrational will not have their way, and that we will be allowed to
live, work, and express ourselves. This is why it is difficult to listen to stories about
violence and suffering. Each of these stories creates a fear in our hearts; it makes
us question humanity, in which we have hope. We start questioning ourselves how
normal and ordinary people can commit such horrible acts. Such stories also make
us aware of our own vulnerability in the world, how easily it could be any of us
who ends up in such terrifying situations.

Sometimes, while visiting survivors, I was at the receiving end of anger and disappointment. These women have been interviewed by many people and over the years they feel abused by the incomers. They feel that everyone benefits from their misery and they get nothing back. Despite their expectations, some women report that talking is not healing, that their situation does not change, and they still don't receive the help they need. They feel that no one really cares about them.

At times I felt that I was paying for other people's mistakes and shortcomings and all I could do was to listen and acknowledge that they were right in feeling angry. There was no use trying to prove to them that I would be different, I could not promise them any form of direct help. All I tried to tell each of my informants was that I would be true to their voice, I would not edit, omit, or change any aspect of their story as some journalists and government officials had done. I would try to tell their story as it is, to include their complaints about the Iraqi government and their disappointment with the Kurdistan Regional Government.

Sometimes when a survivor broke down while telling their story, I too cried with them. I did not know how to console these people and I felt overwhelmed by their pain. Most of the time, however, when I was doing my fieldwork I was getting on with things – organising drivers, choosing villages, finding guides and witnesses, meeting new people all the time and travelling through the Kurdish countryside. The real difficulty started when I returned to Europe and started listening to and transcribing the interviews. I spent many days listening to the testimonies alone and I started having flashbacks and nightmares about the stories I heard.

I experienced Anfal second hand and I started feeling guilty about my own privileged and comfortable life. I also felt responsible towards these women and felt that I had to do something to help. Sometimes this inflated sense of responsibility, this guilt and burden prevented me from doing what I was supposed to do. Many times I have felt that I may not be capable of completing this research. I was too traumatised to work on it. It took me a while to recognise that I was depressed. My friends and family recognised it before I did. Once a friend recommended that I see a therapist but I told him that I didn't need to do that. It is also possible that I suffered from what is known as delayed reaction to migration when years after settling in the new country 'a particular occurrence unexpectedly confronts (the immigrant) with their traumatic past, or echoes the original trauma' (Weinstein and Ortic 1985, cited in Veer 1998: 7).

Over time I became more lethargic and suffered from regular infections and minor illnesses. I remembered these women while eating a good meal, shopping for clothes, watching a film, or just lying down next to my then husband. I kept thinking that these are the basic simple things that have been taken away from these women. I imagined myself and wondered what I would do if that happened to me. Listening to widows talk about their lost husbands made me worry about losing my own. Seeing these women, their wrinkled hands and faces, burnt in the sun and exhausted by hard work, I thought of them and remembered them. I imagined this other life that could have been mine.

I also felt pressurised to make something of the collected data because many survivors had spoken to me despite finding it difficult. Some witnesses said that by talking, they relived the past which they try to forget. But most women talked to me because they thought it was important for these stories to be known. It is important that their suffering is acknowledged. Some talked in the hope that someone, somewhere, will hear their story and will try to help them. This was particularly true in the case of survivors of chemical attacks who desperately need help. It was therefore important that I completed this work.

Eventually, when listening to the testimonies and reading about violence and suffering took a lot out of me, I took time off from this work and sought help. I started seeing a therapist and had regular massage sessions. Encouraged by my husband, I went for long walks and sought help from a nutrition specialist for my health problems. I also took on swimming (something I had always wanted to learn but never quite tackled). Each of these activities helped me a great deal in my process of recovery.

While talking to my therapist, Richard Blackwell, I kept stressing how happy I had been and what a good life I had had before this research started. I had barged through my life, rushing from one language to the next, from one degree to the next, from one project to the next, not once stopping to rest. It was Anfal that eventually brought everything to a halt. My therapist told me something very simple which made me think for a long time. He told me, 'But you are not bulletproof.' He also told me that subconsciously I may have taken this project on because everything else was going well. Maybe I had thought that I could deal with these issues, now that I was feeling strong and happy.

My masseur Ben Pianese once asked me what was wrong in my life because every time I went to see him my muscles were 'rock hard.' This was enough for me to break down. Eventually he told me: 'You don't have to experience pain to understand pain.' He made me see the difference between understanding a problem and becoming a part of it. Later a friend told me a similar thing when she said, 'Remember the purpose of this project, it was to give these women a voice, it was not for you to get depressed and hopeless.'

More recently, while attending a conference in LA, about the role of art in post conflict societies, things became clearer in my mind. Erik Ehn (the Dean of theatre in CalArts) reminded us how violence destroys the safe space to talk and share and how the Arts can provide that missing space for survivors to express themselves. In fact on the first day he said something that really spoke to me. He said: 'If you are responding to communities that have experienced violence it is most likely because there is a ruin inside you that needs tending to. You are never in a complete donor situation.' This really resonates with my experience. I had kept asking myself: Why did I do this research? Why did I become involved in this problem? It took me a while to realise that this problem was my problem, I was not separate from it. It took time to realise that while dealing with the wounds of my community I was in fact addressing my own wounds. The helplessness and suffering expressed by the women reminded me of my own past helplessness. I realised that these problems are a part of what makes me who I am.

Anfal disrupted my peaceful life and clear vision. It took me back to personal issues that I thought I had resolved long ago. I questioned my values, my relationship with my homeland, with my family and friends. I stopped enjoying all things British (Christmas, the understated way people say and do things, the picnics and parties that get going because of alcohol, splitting restaurant bills to the last penny between friends (I have never liked that)). I stopped writing in English and started writing in my mother tongue once again, fell in love with my homeland, felt desperate about how bad things were and felt responsible.

In 2008 I escaped into the abyss of Kurdishness: the dead-end political arguments, the commiserations about bad Kurdish leadership, the loud parties full of (bad) singing and dancing, the friendships that are passionately close but most of the time they end in silence. I took refuge in the warmth of Kurdishness and felt supported by it. Still I could not escape the large questions that had opened up in my life. What do I really want to do in this world? What can I do for my homeland? How can I be happy when there is so much misery in the world, when I am so torn between two such different worlds? How I can I expect my friends and family to relate to all of this?

I think as writers and individuals we keep thinking and re-thinking our position and responsibilities in this world. I have decided to accept my roles, however difficult they may be. I find it difficult to be happy in my personal life (despite having every reason to be) while I am surrounded by injustice and misfortune. I keep thinking of the different kinds of victims in my country – the women, the poor, the traumatised survivors of violence. Recovering from violence is a lengthy process. I can't just retreat into my comfortable life but I also need to protect myself enough to be able to carry on. Learning to balance all of this takes time, it may even take me years.

References

Abdulrahman, Z. 1995. *Death Crematorium: the Anfal Attacks in the Iraqi Regime's Documents*. Tebriz, Iran (Kurdish source).

Ager, A. 1993. *Mental health issues in refugee populations: a review*. Harvard Centre for the Study of Culture and Medicine, Harvard Medical School.

Ala'aldeen, D., Foran, J., House, I., and Hay, A. 1990. Poisoning of Kurdish refugees in Turkey. *The Lancet*, 1990, 335: 287/8.

Aron, A. 1992. Testimonio, a bridge between psychotherapy and sociotherapy, in Cole, E., Espin, O.M. and D-Rothblum, E. (eds) *Refugee women and their mental health: shattered societies, shattered lives*. New York, London, Norwood, Harrington Park Press: 173-189.

Associated Press, 2006. Mass Graves Searched for Second Saddam trial: Evidence allegedly links deposed Iraqi dictator to 1980 Kurdish genocide. 26 June.

Associated Press, 2007. Dutch Businessman who Sold Chemicals to Saddam Appeals War Crimes Conviction. 2 April.

Asukai, N. and Maekawa, K. 2002. Psychological and physical health effects of the 1995 Sarin attack in the Tokyo subway, in Havenaar, J.M., Cwikel, J.G., and Bromet, E.J. (eds) *Toxic turmoil: Psychological and societal consequences of ecological disasters*. The Plenum series on stress and coping. Series editor, Donald Meihenbaum: 149-162.

Baban, T. 2000. *Press Conference at the United Nations Foreign Press Centre* (New York). September 26.

Balali-Mood, M., Mousavi, S.H., Balali-Mood, B. 2008. Chronic health effects of sulphur mustard exposure with special reference to Iranian veterans. *Emerging Health Threats Journal* 2008, 1:e7.

Baron, N., Jenson, S.B., and de Jong, J.T.V.M. 2003. Refugees and internally displaced people, in Green, B.L., Friedman, M.J., de Jong, J.T.V.M., Solomon, S.D., Keane, T.M., Fairbank, J.A., Donelan, B., Frey-Wouters, E. and Danieli, Y., *Trauma intervention in war and peace: Prevention, practice and policy*. Springer: 243-265.

Bassuk, E.L., Donelan, B., Selema, B., Ali, S., de Aguiar, A.C., Eisenstein, E., Vostanis, P., Varavikova, E., and Tashjian, M. 2003. Social Deprivation in Green, B.L., Friedman, M.J., de Jong, J.T.V.M., Solomon, S.D., Keane, T.M., Fairbank, J.A., Donelan, B., Frey-Wouters, E. and Danieli, Y., *Trauma intervention in war and peace: Prevention, practice and policy*. Springer: 35-55.

BBC. 2005. Killing of Iraq Kurds 'genocide.' 23 December.

BBC. 2006. Saddam Trial Hears of gas attacks. 22 August.

Beaton, R. and Murphy, S. 2002. Psychosocial Responses to Biological and Chemical Terrorist Threats and Events: Implications for the Workplace.

Journal of the American Association of Occupational Health Nurses, 50(4): 182-189.

Bello, D., and Doucet, J. 2003. Poverty, inequality, and discrimination as sources of depression among U.S. women, *Psychology of Women Quarterly*, 27 (2003): 101-113. Blackwell Publishing. Printed in the USA.

Biggam, F.H. and Power, K.G. 1997. Social support and psychological distress in a group of incarcerated young offenders. *International Journal of Offender Therapy and Comparative Criminology* 41(3): 213-230.

Blas, E., Gilson, L., Kelly, M.P., Labonté, R., Lapitan, J., Muntaner, C., Östlin, P., Popay, J., Sadana, R., Sen, G., Schrecker, T., and Vaghri, Z. 2008. Addressing social determinants of health inequities: what can the state and civil society do? *Lancet* 2008; 372: 1684-89.

Bondy, R. 1998. Women in Theresienstadt and the Family Camp in Birkenau, in Ofer, D. and Weitzman, L.J. (eds) *Women in the Holocaust*. New Haven and London: Yale University Press.

Bos, P.R. 2003. Women and the Holocaust: Analyzing gender difference, in Baer, E.B. and Goldenberg, M. (eds) *Experience and expression: Women, the Nazis and the Holocaust*.

Brown, L.S. 2004. Feminist paradigms of trauma treatment. *Psychotherapy: Theory, Research, Practice, Training* 2004, 41(4): 464-471.

Burstow, B. 1992. *Radical Feminist Therapy: Working in the context of violence.* London: Sage Publications.

Butalia, U. 1997. A question of silence: Partition, women and the state, in Lentin, R. (ed) *Gender and catastrophe*. London and New York: Zed Books.

Chaliand, G. 1993. *A people without a country: The Kurds and Kurdistan.* London: Zed Books.

Danieli, Y. 2007. Essential elements in healing from massive trauma: some theory, victims' voices, and international developments, in Miller, J. and Kumar, R. (eds), *Reparations: Interdisciplinary inquiries*. Oxford, United Kingdom: Oxford University Press: 307-322.

Daragahi, B. 2006. Buried but not nameless in Iraq's desert: Victims' hidden ID cards may assist Hussein trial. *Los Angeles Times*. June 29.

Das, V. 1997. Language and body: transactions in the construction of pain, in Kleinman, A., Das, V., and Lock, M. (eds) *Social Suffering*. Berkeley, Los Angeles, London: University of California Press.

Dasgupta, S.D. 1998. Gender roles and cultural continuity in the Asian Indian immigrant community in the USA. *Sex Roles*. 38(11/12): 953-655.

DeChant, B. 1996. *Women and group psychotherapy: Theory and practice*. New York, NY, USA: The Guilford Press.

Distcheid, C., July 2003. Women speak out against dishonorouable crimes. *Refugee women's news*. Issue 23. London. Refugee Women's Association.

Egan, D.G. and Perry, S.K. 2001. Gender identity: a multidimensional analysis with implications for psychosocial adjustment. *Developmental Psychology*. 37(4): 451-463.

Elmadmad, K. 1999. The human rights of refugees with special reference to Muslim refugee women, in Indra, D. (ed), *Engendering Forced Migration. Refugee and forced migration studies*. Volume 5. New York. Oxford: Berghahn Books.

Espin, O.M. 1992. Roots uprooted: The psychological impact of historical/ political dislocation, in Cole, E., Espin, O.M., and D-Rothblum, E. *Refugee women and their mental health: shattered societies, shattered lives*. New York, London, Norwood, Harrington Park Press: 9-20.

Espin, O.M. 1996. 'Race', racism, and sexuality in the life narratives of immigrant women, in Wilkinson, S., *Feminist Social Psychologies: International Perspectives*. Scarborough, North Yorkshire, Open University Press: 87-103.

Fein, H. 1993. *Genocide, a sociological perspective*. London, Newbury Park, New Delhi: Sage Publications.

Ferguson, E. 2001. Personality and coping traits. *British Journal of Health Psychology*. 6: 311-325.

Ghazanfari, T., Faghihzadeh, S., Aragizadeh, H., Soroush, M.R., Yaraee, R., Hassan, Z.M., Foroutan, A., Vaez-Mahdavi, M.R., Javadi, M.A., Moaiedmohseni, S., Azizi, F., Panahi, Y., Mostafaie, A., Ghasemi, H., Shams, J., Pourfarzam, S., Jalali-Nadoushan, M.R., Fallahi, F., Ebtekar, M., Davoudi, S.M., Ghazanfari, Z., Ardestani, S.K., Shariat-Panahi, S., Moin, A., Rezaei, A., Kariminia, A., Ajdary, S., Mahmoudi, M., Roshan, R., Ghaderi, S., Babai, M., Naghizadeh, M.M., Ghanei, M. 2009 Sardasht-Iran Cohort Study of Chemical Warfare Victims: Design and Methods. *Archives of Iranian Medicine* 2009; 12(1): 5-14.

Garwood, A. 2002. The holocaust and the power of powerlessness: survivor guilt an unhealed wound, in Covington, C., Williams, P., Arundale, J., and Knox, J. (eds) *Terrorism and war: unconscious dynamics of political violence*. London: Karnac Books.

Goldenberg, M. 1998. Memories of Auschwitz survivors: the burden of gender, in Ofer, D. and Weitzman, L.J. (eds) *Women in the Holocaust*. New Haven and London: Yale University Press.

Gosden, C. 1998a. Why I Went, What I Saw. *The Washington Post*, Wednesday 11 March 1988. Page A19.

Gosden, C. 1998b. *Chemical and Biological Weapons Threats to America: Are we prepared?* Testimony of Dr Christine M. Gosden before the Senate Judiciary Subcommittee on Technology, Terrorism and Government and the Senate Select Committee on Intelligence. April 22nd 1998.

Gosden, C. and Gardener, D. 2002. Lesson of Iraq's mass murder. *Washington Post*. June 2 2002.

Grant-Pearce, C. and J. Deane 1999. Joint working between the public and purchasing authorities to determine mental health information needs, in Bhugra, D. and Bahl, V. (eds) *Ethnicity: An Agenda for Mental Health*. London: The Royal College of Psychiatrists: 166-173.

Green, B.L. 2003. Traumatic stress and its consequences, in Green, B.L., Friedman, M.J., de Jong, J.T.V.M., Solomon, S.D., Keane, T.M., Fairbank, J.A., Donelan,

B., Frey-Wouters, E. and Danieli, Y., *Trauma intervention in war and peace: Prevention, practice and policy.* Springer: 17-32.

Grieco, M. 2006. Hagerstrand, hegemony and distributed discourse: the use of the world wide web in tracking contemporary migration paths. *European Spatial Research and Policy*, 13(1).

Hashemian, F., Khoshnood, K., Desai, M. M., Falahati, F., Kasl, S., and Southwick, S. 2006. Anxiety, depression, and Post traumatic stress in Iranian survivors of Chemical Warfare. *Journal of the American Medical Association*, 296: 560-566.

Herbst, A. 1992. From helpless victim to empowered survivor: oral history as a treatment for survivors of torture, in Cole, E., Espin, O.M. and D-Rothblum, E. (eds) *Refugee women and their mental health: shattered societies, shattered lives.* New York, London, Norwood, Harrington Park Press.

Herman, J.L. 1997. *Trauma and recovery.* New York: Basic Books.

Heyndrickx, A., 27/4/1988. *Clinical Toxicological Reports and Conclusion of the biological samples of men and the environmental samples brought to the department of Toxicology and the State University of Ghent for Toxicological Investigation.* Report no. 88/KU2/PJ881, Department of Toxicology, University of Ghent.

Hilberg, R., 1985. *The destruction of the European Jews.* New York: Holmes and Meier.

Hiltermann, J. Summer 2000. *Elusive justice: trying to try Saddam.* Middle East Watch Report, no. 215.

Hiltermann, J. 2007. *A poisonous affair: America, Iraq, and the gassing of Halabja.* Cambridge University Press.

Hiroto, D.S. and Seligman, M.E. P. 1975. Generality of learned helplessness in man. *Journal of Personality and Social Psychology*, 31, 311-327.

International Centre for Transitional Justice 2007a. *The Anfal trial and the Iraqi High Tribunal, Update Number Three: The Defense Phase and the Closing Stages of the Anfal Trial.* 24 June http://www.ictj.org/images/content/7/2/726. pdf.

International Centre for Transitional Justice 2007b. *Iraqi Tribunal issues verdict in Anfal case.* 24 June. http://www.ictj.org/en/news/press/release/1240.html.

Jack, D. 1991. *Silencing the Self: Women and Depression.* Cambridge: Harvard University Press.

Jack, D. and Dill, D.C. 1992. The silencing the self scale; schemas of intimacy associated with depression in women. *Psychology of Women Quarterly.* 16: 97-106.

Kim, G., Torbay, R., and Lawry, L. 2007. Basic Health, Women's Health, and Mental Health Among Internally Displaced Persons in Nyala Province, South Darfur, Sudan. *American Journal of Public Health*, February 2007, 97(2).

Kluegel, J.R., and Smith, E.R. 1986. *Beliefs about inequality:Americans' views of what is and what ought to be.* New York: Aldine De Gruyter.

Knickmeyer, E. 2005. 113 Kurds are found in mass graves: Hussein victims almost all women, children. *Washington Post Foreign Service*, Saturday, April 30, Page A09.

Kulwicki, A.D. 2002. The practice of honour killing crimes: a glimpse of domestic violence in the Arab world. *Issues in mental health and nursing*. 23: 77-87.

Kuper, L. 1981. *Genocide, its political use in the twentieth century*. Harmondsworth, UK: Penguin Books.

Langer, L. 1991. *Holocaust testimonies: the ruins of memory*. Yale University Press. New Haven and London.

Langer, L. 1998. Gendered Suffering? Women in Holocaust testimonies, in Ofer, D. and Weitzman, L.J. (eds) *Women in the Holocaust*. New Haven and London: Yale University Press.

Lengua, L.J. and Stormshak, E. 2000. Gender, gender roles and personality: Gender differences in the prediction of coping and psychological symptoms. *Sex Roles*, 43: 787-820.

Lengyel, O. 1993. Scientific experiments, in Rittner, C. and Roth, J.K. (eds) *Different voices: women and the Holocaust*. St. Paul, Minnesota: Paragon House.

Leydesdorff, S., Passerini, L., and Thompson, P. 1996. International Yearbook of Oral History and Life stories, Volume IV. *Gender and Memory*. Oxford: Oxford University Press.

Light, D. 1992. Healing their wounds: Guatemalan refugee women as political activists, in Cole, E., Espin, O.M. and D-Rothblum, E. (eds) *Refugee women and their mental health: shattered societies, shattered lives*. New York, London, Norwood, Harrington Park Press: 297-308.

Lynch, C. 2005. Oil-for-Food Panel Rebukes Annan, Cites corruption. *Washington Post Staff Writer*, 8 September.

MacAskill, E. 2005. Oil-for-Food Report Condemns 'Corrupt' UN. *The Guardian*, 7 September.

Makiya, K. 1992. The Anfal: Uncovering an Iraqi campaign to exterminate the Kurds. *Harper's Magazine*, May 1992: 53-61.

Makiya, K. 1993. *Cruelty and Silence: war, tyranny, uprising and the Arab world*. W.W. Norton and Company. New York and London.

Mastussek, P., Grigat, R., Haibock, H., Halbach, G., Kemmler, R., Mantell, D., Triebel, A., Vardy, M., Wedel, G. 1975. *Internment in concentration camps and its consequences*. Springer-Verlag New York, Heidelberg, Berlin.

McDowall, D., 2004. *A modern history of the Kurds*. Third Edition. I.B. Tauris. London and New York.

Middle East Watch 1993. *Genocide in Iraq: The Anfal campaign against the Kurds*. New York, Washington, Los Angeles, London: Human Rights Watch.

Middle East Watch and Physicians for Human Rights 1993. *The Anfal campaign in Iraqi Kurdistan: The destruction of Koreme*. New York, Washington, Los Angeles, London: Human Rights Watch.

Mlodoch, K. 2008. *Violence, memory and dealing with the past in Iraq: The perspective of Anfal survivors in Kurdistan* (conference paper). Writing the history of Iraq: Historiographical and Political challenges.

Mlodoch, K. 2009. *We want to be remembered as strong women and not as shepherds – Anfal surviving women in Kurdistan-Iraq struggling for agency and acknowledgement* (workshop paper). Gendered Memories: cultural representations, local practices, and transnational regimes in the Middle East and North Africa. Zentrum Moderner Orient, Berlin, 11-12 June 2009.

Myers, F., McCollam, A., Woodhouse, A. 2005. Addressing mental health inequalities in Scotland: equal minds, working paper summary. National program for improving mental health and wellbeing. *Scottish development centre for mental health.*

Ofer, D., and Weitzman, J. 1998. *Women in the Holocaust.* New Haven and London: Yale University Press.

Omar, A. 2003. *Kurdish women and Anfal* (Kurdish source). Kurdistan: Aras Publishing (Kurdish source).

Omar, A. 2007. *The Anfal campaign and its consequences on financial, psychological and social situation of women.* Conference paper. Erbil, 26/1/08 (Kurdish source).

Physicians for Human Rights 1989. *Winds of death: Iraq's use of poison gas against the Kurdish population.* Report of a medical mission to Turkish Kurdistan. Physicians for Human Rights.

Postero, N.G. 1992. On trial in the promised land: seeking asylum, in Cole, E., Espin, O.M. and D-Rothblum, E. (eds) *Refugee women and their mental health: shattered societies, shattered lives.* New York, London, Norwood, Harrington Park Press: 173-189.

Power, S. 2002. A problem from hell: America and the age of Genocide. New York: Basic Books.

Quirk, G.J., Casco, L. 1994. Stress disorders of families of the disappeared: A controlled Study in Honduras. *Social Sciences and Medicine,* 39(12): S.1675 -1679. London, 1994.

Qurbany, A. 2003a. *Anfal witnesses.* Suleimanya: Kurdistan (Kurdish source).

Qurbany, A. 2003b. *Anfal witnesses.* Chamchamal: Kurdistan (Kurdish source).

Qurbany, A. 2004. *From Um Re'an to Topzawa: the bulldozer driver who covered some of the Anfal victims.* Kirkuk: Tishk Publishing House. (Kurdish source).

Rampage, C. 1991. Personal authority and women's self-stories. Goodrich, T. J., *Women and Power: Perspectives for family therapy.* New York, London, WW. Norton & Company: 109-122.

Resool, S. 1990. *Destruction of a nation.* USA: Harith Zahawi and Latif Rashid.

Resool, S. 2003. *Anfal: The Kurds and the Iraqi state.* London (Kurdish source).

Reterska, K. and Sissons, M. 2007. Iraq tribunal issues verdict in Anfal case: Tribunal's flaws persist in historic trial for Iraqi Kurds. *International Centre for Transitional Justice,* For immediate release, 24 June.

Rewan. A woman gets killed by her own brother. *Rewan.* No.162, 16 January.

Ringelheim, J. 1998. Genocide and gender: A split memory, in Ofer, D. and Weitzman, L.J. (eds) *Women in the Holocaust*. New Haven and London: Yale University Press.

Roberts, B., Felix Ocaka, K., Browne, J., Oyok, T. and Sondorp, E. 2008. Factors associated with the health status of internally displaced persons in northern Uganda. *Journal of Epidemiol Community Health*. 63: 227-232.

Roe, M.D. 1992. Displaced women in settings of continuing armed conflict, in Cole, E., Espin, O.M. and D-Rothblum, E. (eds) *Refugee women and their mental health: shattered societies, shattered lives*. New York, London, Norwood, Harrington Park Press: 173-189.

Romano, J.A., and King, J.M. 2001. Psychological factors in chemical warfare and terrorism, in Somani, S.M. and Romano, J.A. (eds) *Chemical warfare agents: toxicity at low level*. CRC Press LLC.

Romano J.A., Lumley, L.M., King, J.M., and Saviolakis, G.A. 2007. Chemical warfare, chemical terrorism and traumatic stress response: An assessment of psychological impact, in Romano, J.A., Lukey, B.L., Salem, H. (eds) *Chemical Warfare Agents: Pharmacology, Toxicology, and Therapeutics*. Second Edition. CRC Press. Taylor and Francis Group.

Ross, F. 2001. Speech and Silence: Women's Testimonies in the First Five Weeks of Public Hearings of the South African Truth and Reconciliation Commission, in Das, V., Kleinman, A., Lock, M., Ramphele, M., and Reynolds, P. (eds) *Remaking a World: Violence, Social Suffering and Recovery*. Berkeley, Los Angelese, London: University of California Press.

Roy, A. 2001. *Power Politics*. South End Press.

Saldana, D.H. 1992. Coping with stress: a refugee's story, in Cole, E., Espin, O.M. and D-Rothblum, E. (eds) *Refugee women and their mental health: shattered societies, shattered lives*. New York, London, Norwood, Harrington Park Press.

Segal, P. 2009. Resource Rents, Redistribution, and Halving Global Poverty: The Resource Dividend. *Oxford Institute for Energy Studies*, Working Paper SP 22. http://www.oxfordenergy.org/pdfs/SP22.pdf.

Sen, A. 1999. *Development as freedom*. Oxford University Press.

Seshadri, K.R. 2008. When Home Is a Camp, Global Sovereignty, Biopolitics, and Internally Displaced Persons. *Social Text 94*. 26(1).

Silove, D., Steel, Z. and Watters, C. 2000. Policies of Deterrence and the Mental Health of Asylum Seekers. *Journal of the American Medical Association* 284(5): 604-611.

Solomon, S.D. 2003. Introduction, in Green, B.L., Friedman, M.J., de Jong, J.T.V.M., Solomon, S.D., Keane, T.M., Fairbank, J.A., Donelan, B., Frey-Wouters, E. and Danieli, Y., *Trauma intervention in war and peace: Prevention, practice and policy*. Springer.

Stein, B. 1986. The experience of being a refugee: Insights from the research literature, in Williams, C.J. and Westermeyer, J. (eds) *Refugee mental health in resettlement countries*. Washington: Hemisphere Publishing Corporation.

Stokes, J.W., Banderet, L.E. 1997. *Psychological Aspects of Chemical Defense and Warfare. Military Psychology.* 9(4): 395-415.

Summerfield, D. 1999. A critique of seven assumptions behind psychological trauma programmes in war-affected areas. *Social Science and Medicine,* 48, 1449-1462.

Swan, V. 1998. Narrative therapy, feminism, and race, in Seu, I.B. and Heenan, C. (eds) *Feminism and psychotherapy: Reflections on contemporary theories and practices.* London, Thousand Oaks, New Delhi, SAGE Publications: 30-42.

Tageszeitung 2002. *Kofi Annan übt leise Selbstkritik.* taz Nr. 6928 vom 12.12.2002, Seite 10, 78 TAZ-Bericht B. S.

Tec, N. 2003. *Resilience and Courage: women, men and the Holocaust.* Yale University Press. New Haven and London.

Todeschini, M. 2001. The Bomb's Womb? Women and the Atom Bomb, in Das, V., Kleinman, A., Lock, M., Ramphele, M., and Reynolds, P. (eds) *Remaking a world: violence, social suffering, and recovery.* Berkeley and Los Angeles, California: University of California Press.

Turner, S., Yuksel, S. and Silove, D. 2003. Survivors of mass violence and torture, in Green, B.L., Friedman, M.J., de Jong, J.T.V.M., Solomon, S.D., Keane, T.M., Fairbank, J.A., Donelan, B., Frey-Wouters, E. and Danieli, Y. (eds) *Trauma intervention in war and peace: Prevention, practice and policy.* Springer: 185-211.

United Nations 2003. *Iraqi Oil Sales Fund Humanitarian Action.* Office of the Iraq Programme, 21 November. http://www.un.org/Depts/oip.

United Nations 2009. *Oil for Food. Office of the Iraq Programme,* 10 October. http://www.un.org/Depts/oip/background/index.html.

United States Senate. October 1988. *Chemical Weapons use in Kurdistan: Iraq's final offensive.* A staff Report to the Committee on Foreign Relations.

Vago, L.R. 1998. One year in the black hole of our planet earth, a personal narrative in Ofer, D. and Weitzman, L.J. (eds) *Women in the Holocaust.* New Haven and London, Yale University Press: 273-284.

Van Bruinessen, M. 1997. *Kurdistan in the shadow of history.* Susan Meiselas. New York, Random House.

Veer, G.V.D. 1998. *Counselling and therpay with refugees and victims of torture.* Chichester. New York, Weinheim, Brisbane, Singapore, Toronto, John Wiley & Sons: 0-471-98226-1.

Watters, C. 2001. Emerging paradigms in the mental health care of refugees. *Social Science and Medicine.* 52: 1709-1718.

Watters, C. 2002. *Asylum seekers and mental health care in the UK. Breathings space project.* Tracking Report.

Weber, L. 1998. A conceptual framework for understanding race, class, gender and sexuality. *Psychology of Women Quarterly.* 22: 13-32.

Weiss, T.G. 1999. Whither International Efforts for Internally Displaced Persons? *Journal of Peace Research* 36: 363-73.

Williams, J. 2004. Women's mental health – taking inequality into account, in Tew, J. (ed) *Social perspectives on mental health: developing social models to understand and work with mental distress.* London: Jessica Kingsley.

World Health Organisation 2000. *Women's Mental Health, an evidence based review.* Geneva. World Health Organisation.

World Health Organisation 2009. *Mental Health, resilience and inequality.* WHO Regional Office for Europe.

Zumach, A. 2002. Deutsche Hilfe für *Bagdad:* Bericht an UN-Sicherheitsrat enthält Namen von über 80 deutschen Firmen, die Saddam Hussein bei Rüstungsgeschäften unterstützten. USA suchen weitere Hinweise auf schmutzige Geschäfte. *Die Tageszeitung.* 17 December, taz Seite 1 67 Zeilen, S1.

Index

Printed in the United States
by Baker & Taylor Publisher Services